FROM GRAY TO GREEN:
An Urban Forest Master Plan for East Palo Alto

PREPARED FOR
The City of East Palo Alto
Adopted April 19, 2022

PREPARED BY
San Francisco Estuary Institute
and **HortScience | Bartlett Consulting**

SFEI San Francisco Estuary Institute

HORT SCIENCE
BARTLETT CONSULTING

AUTHORS
Megan Wheeler (SFEI)
Lauren Stoneburner (SFEI)
Erica Spotswood (SFEI)
Robin Grossinger (SFEI)
Darya Barar (HortScience)
Carol Randisi (HortScience)

SPATIAL ANALYSIS and CARTOGRAPHY
Lauren Stoneburner (SFEI)

REPORT DESIGN
Ellen Plane (SFEI)
Ruth Askevold (SFEI)

FUNDED BY California Department of Forestry and Fire Protection, California Climate Investments Program
SFEI-ASC Publication #1071

SUGGESTED CITATION

San Francisco Estuary Institute and Bartlett Consulting. 2022. From Gray to Green: An Urban Forest Master Plan for East Palo Alto. Prepared for the City of Palo Alto. A report of SFEI-ASC's Resilient Landscapes Program, Publication #1071, San Francisco Estuary Institute, Richmond, CA.

REPORT AVAILABILITY

Report is available on SFEI's website at www.sfei.org/projects/east-palo-alto-urban-forest-master-plan

Text underlined in blue in the printed copies of this report indicates that a web link is available. To access those links, please refer to the digital PDF version of the report (www.sfei.org/projects/east-palo-alto-urban-forest-master-plan).

COVER CREDITS

East Palo Alto, view down University Ave toward the Dumbarton Bridge. (Cover image courtesy of Droneshot: droneshot.com)

CONTENTS

Section 1: Introduction and Context

Section 2: Current Status

Section 3: Plan for the Future

Appendices (AVAILABLE ONLINE*)

* VISIT HTTPS://WWW.CITYOFEPA.ORG/ECONDEV/PAGE/URBAN-FOREST-MASTER-PLAN

ACKNOWLEDGEMENTS

Funding for the development of this Plan was provided by the California Department of Forestry and Fire Protection (Cal Fire) as part of the California Climate Investments Program. Additional funding was provided by the Sand Hill Foundation and the Robert Wood Johnson Foundation, to support additional engagement with the community and evaluation of the public health implications of expanding the urban forest.

The *From Gray to Green: An Urban Forest Master Plan for East Palo Alto* Project is part of California Climate Investments, a statewide program that puts billions of Cap-and-Trade dollars to work reducing greenhouse gas emissions, strengthening the economy, and improving public health and the environment– particularly in disadvantaged communities. The Cap-and-Trade program also creates a financial incentive for industries to invest in clean technologies and develop innovative ways to reduce pollution. California Climate Investments projects include affordable housing, renewable energy, public transportation, zero-emission vehicles, environmental restoration, more sustainable agriculture, recycling, and much more. At least 35 percent of these investments are located within and benefiting residents of disadvantaged communities, low-income communities, and low-income households across California. For more information, visit the California Climate Investments website at: www. caclimateinvestments.ca.gov.

We are grateful for feedback on ideas and drafts of this plan from the members of our community steering committee: Bronwyn Alexander, Najiha Al Asmar, Maya Briones, Robert Allen Fisk, Lileiti Grew, Elizabeth Jackson, Antonio Lopez, Catherine Martineau, Laura Martinez, Jeff Poetsch, and Violet Saena. Special thanks to Nuestra Casa and Anamatongi Polynesian Voices for hosting a focus group meeting, and to all community members who took the time to speak with us or submit survey responses as part of our community input gathering. Canopy donated a fruit tree as a raffle prize, in addition to providing valuable insight into planting opportunities and constraints in the city. Photos for this report were provided courtesy of Canopy and Pro Bono Photography.

Many people contributed to the text and ideas in this plan. Thanks to Catherine Martineau, Uriel Hernandez, Michelle Daher, and Jimi Schied for the early conceptualization of this plan and for their significant contributions along the way. Thanks to East Palo Alto staff members who contributed their knowledge and ideas: Amy Chen, Elena Lee, Humza Javed, Jay Farr and Patrick Heisinger. Additional writing, review, design, and background research was done by Clara Kieschnick, Crissy Pickett, Erik Ndayishimiye, Stephanie Panlasigui, Vanessa Lee, Ellen Plane, Brandon Herman, and Jennifer Symonds.

The Urban Forest Master Plan for East Palo Alto is a comprehensive and timely guide to expand and manage our tree landscape. In the past, intensive agriculture and careless urban development devastated the population of trees in what is now the City of East Palo Alto. Restoring nature and repopulating an urban forest is of paramount importance to the overall health and well being of the community. Let's all do our part now and in the future to follow this plan.

– East Palo Alto Mayor **Ruben Abrica**

VISION STATEMENT

EAST PALO ALTO

is a diverse community with strong roots
in land stewardship, connection to nature,
and a commitment to building
the urban canopy.

East Palo Alto's urban forest will include a
diverse mix of healthy trees that provide
benefits to residents across the city
by reducing air pollution, heat, noise,
flooding, and stress.

An expanded urban forest will close the
"Green Gap" and bring East Palo Alto the
mental, physical, and ecological health
benefits that surrounding
communities experience.

EXECUTIVE SUMMARY

Introduction and context

The urban forest is an important part of urban infrastructure. Trees can provide shade, help mitigate air pollution, create beautiful and relaxing places, reduce stormwater runoff, store carbon, provide fruit, and support wildlife. In order to ensure that the community of East Palo Alto can enjoy the benefits provided by trees, it is necessary for the City to invest in their planting, management, and protection. This Urban Forest Master Plan is intended to help the City effectively manage this important resource to achieve benefits for community members.

East Palo Alto historically included areas of willow grove and oak woodland, both landscapes containing trees. These landscapes were converted to agriculture and then suburban development, though some native oaks remain. The city faces numerous environmental challenges in the future, including increased flood and heat risk, health disparities, and limited potable water resources. Planning for a robust urban forest can help to address some of these concerns, and should take into account both current and future conditions.

Current status of the urban forest

The current urban forest was evaluated by quantifying the existing tree canopy cover in the city and analyzing the most recent public tree inventory conducted. About 13.5% of East Palo Alto is currently covered by tree canopies, far below canopy cover in neighboring Menlo Park (27%) or Palo Alto (38%). Histories of inequitable investment and development likely account for these differences, which also impact environmental and health outcomes.

The City manages 5,745 public trees representing 253 different species. 15% of public trees are California native species, which are important for supporting local birds and other wildlife. Many of the trees are still small, reflecting recent planting efforts. Public trees provide numerous benefits to residents, including carbon sequestration and storage, air pollution removal, wildlife support, creating beautiful places, and a wide variety of mental and physical health benefits.

Trees in East Palo Alto are currently regulated by three ordinances within the municipal code. Any street tree or tree in a public place within the city, and any private tree 24" or greater in circumference is protected and requires a permit for removal. Management authority and procedures regarding street trees are not clearly defined in the existing ordinances. Most tree planting in the city is done by the nonprofit Canopy. About $150,000 is budgeted annually by the City for public tree management, which is largely carried out by a contractor in response to reported problems. No arborist is employed on staff. Recommended management practices include a regularly updated tree inventory to enable a proactive maintenance approach, which should be clearly defined in city ordinances along with tree protection measures.

During development of this Urban Forest Master Plan, the team engaged with over 350 community members to understand opportunities and barriers for expanding the urban tree canopy. Community members expressed positive opinions of trees overall and a desire to grow the urban forest, along with concerns about appropriate management and maintenance, vulnerability to climate change and water restrictions, and application of the current tree protection ordinance.

Plan for the future

Following these analyses and in consultation with the project community steering committee, two goals were identified to enhance the City's urban forestry program. *The table on the following pages details objectives toward these goals, along with actions the City can take to achieve each objective.*

From left: Former Mayor Laura Martinez, Mayor Ruben Abrica, Teen Urban Foresters, and partners. Bayshore Christian Ministries tree planting in the fall of 2019. Photo: Canopy.

Goal 1: Grow a healthy, extensive, vibrant, and diverse urban forest to provide 20% canopy cover by 2062 and 30% by 2122.

Objective	Action	Key Point
1. Grow the urban forest.	1. Develop a city-wide tree planting strategy to increase equity within East Palo Alto and with neighboring cities.	1. Plant 200-250 public trees per year
		2. Plant 150-200 private trees per year
		3. Measure city-wide and zone-specific tree canopy cover every 10 years to track progress
	2. Plant climate-adapted species with the intention of planting the right tree in the right place to maximize benefits for public health, wildlife, and resilience.	1. Plant appropriate trees for each habitat zone in the city
		2. Plant the right species for the specific planting location
		3. Maintain species and age diversity within the urban forest
	3. Integrate trees into future urban design.	1. Set standard sizes for tree wells and planting strips to allow for healthy tree growth
		2. Enact a parking lot shade ordinance
		3. Review development plans for tree protection and planting
2. Define responsibilities and support improved maintenance practices and protections for public and private trees.	4. Codify responsibility and set standards for proactive public tree maintenance and protection.	1. Codify the City's authority and responsibility for street tree maintenance
		2. Take a proactive grid pruning approach to public tree maintenance
		3. Condense the City code regulating public trees into a single ordinance with all key components, including standards of care for public trees
	5. Revise the tree protection ordinance and implementation process to provide strong protection for private trees.	1. Revise the definition of a protected tree
		2. Revise the criteria for protected tree removal
		3. Require development projects to apply for tree removal permits
		4. Increase staff capacity to manage the tree protection ordinance by hiring or contracting with a certified arborist
		5. Review and clarify fees and requirements for tree removal permit applications
		6. Create a more robust appeals process with community input
		7. Publicize and enforce tree protections
	6. Seek opportunities for additional funding.	1. Identify opportunities to obtain additional funding to support urban forestry programs

Goal 2: Connect with an engaged and informed community to provide stewardship of the urban forest.

Objective	Action	Key Point
3. Connect with the community around tree stewardship.	7. Design a cohesive and inclusive public outreach program focused on building awareness of the benefits of trees and how and why trees are protected in the city.	1. Engage with local community groups in urban forest stewardship activities
		2. Partner with Canopy to identify opportunities to expand current community outreach programs
		3. Provide up-to-date information about tree protections and management on the City website
		4. Translate tree information into common non-English languages spoken in East Palo Alto
		5. Designate a forum for the public to engage with tree management and protection actions
		6. Identify ways for the City to celebrate trees
	8. Become a Tree City USA.	1. As in Action 4, update the public tree ordinance to set standards of care
		2. As in Actions 5 and 7, designate or create a public body to review tree removal applications and appeals, and to serve as a public forum for tree-related issues
		3. As in Action 7, plan celebratory activities for Arbor Day

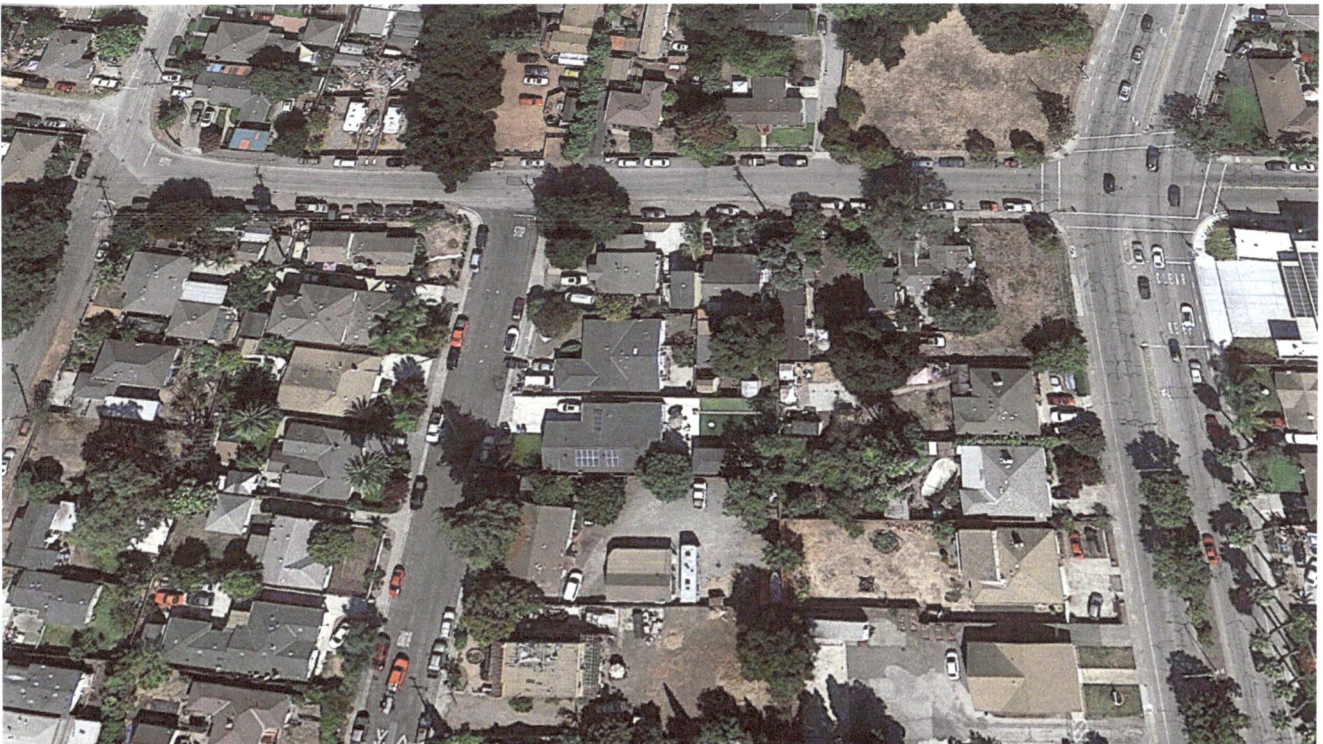

East Palo Alto. Imagery: Google Earth.

DECLARACIÓN DE VISIÓN

EAST PALO ALTO

es una comunidad diversa con fuertes raíces en la administración de la tierra, conexión con la naturaleza y dedicación a construir un bosque urbano.

El bosque urbano de East Palo Alto incluirá una mezcla diversa de árboles saludables que brindan beneficios a los residentes de toda la ciudad al reducir la contaminación del aire, el calor, el ruido, las inundaciones y el estrés.

Un bosque urbano expandido cerrará la *"Brecha Verde"* y traerá a East Palo Alto los beneficios para la salud mental, física y ecológica que experimentan las comunidades circundantes.

Resumen ejecutivo

Introducción y contexto

El bosque urbano es una parte importante de la infraestructura urbana. Los árboles pueden dar sombra, ayudar a mitigar la contaminación del aire, crear lugares hermosos y relajantes, reducir la escorrentía de aguas pluviales, almacenar carbono, proveer fruta y apoyar la vida silvestre. Para garantizar que la comunidad de East Palo Alto pueda disfrutar de los beneficios proporcionados por los árboles, es necesario que la Ciudad invierta en su plantación, manejo y protección. Este Plan Maestro de Bosques Urbanos está destinado a ayudar a la Ciudad a administrar de manera efectiva este importante recurso para lograr beneficios para miembros de la comunidad.

East Palo Alto históricamente incluía áreas de arboledas saucos y bosques de robles, ambos paisajes que contenían árboles. Estos paisajes se convirtieron en agricultura y luego en desarrollo suburbano, aunque quedan algunos encinos nativos. La ciudad enfrenta numerosos retos ambientales en el futuro, incluyendo el aumento del riesgo de inundaciones y calor, disparidades de salud y recurso limitados de agua potable. La planificación de un bosque urbano robusto puede ayudar a dirigir algunas de estas preocupaciones, y debe tener en cuenta las condiciones actuales y futuras.

Estado actual del bosque urbano

El bosque urbano actual se evaluó cuantificando la cobertura árborea existente en la ciudad y analizando el inventario público de árboles más reciente realizado. Cerca de 13.5% de East Palo Alto está actualmente cubierto por copas árboreas, mucho menos que la copa árborea de Menlo Park (27%) o Palo Alto (38%). Es probable que historias de inversión y desarrollo inequitativos explican estas diferencias, que también afectan los resultados ambientales y de salud.

La Ciudad administra 5,745 árboles públicos que representan 253 especies diferentes. El 15% de árboles públicos son especies nativas de California, que son importantes para apoyar las aves locales y otros animales salvajes. Muchos de los árboles todavía son pequeños, lo que refleja los recientes esfuerzos de plantación. Los árboles públicos proporcionan numerosos beneficios a los residentes, la eliminación de la contaminación del aire, el apoyo a la vida silvestre, la creación de lugares hermosos y una amplia variedad de beneficios para la salud mental y física.

Los árboles en East Palo Alto están actualmente regulados por tres ordenanzas dentro del código municipal. Cualquier árbol en la calle o árbol en un lugar público dentro de la ciudad, y cualquier árbol privado de 24 "o más de circunferencia está protegido y requiere un permiso para removerlo. La autoridad de Administración y procedimientos relativos a árboles en la calle no están claramente definidos en las ordenanzas existentes. La mayor parte de la plantación de árboles en la ciudad es realizada por la organización sin fines de lucro Canopy. Alrededor de $150,000 son presupuestados anualmente por la Ciudad para el manejo público de árboles, que es llevado a cabo en gran parte por un contratista en respuesta a los problemas reportados. No se emplea ningún arborista en el personal. Las prácticas de manejo recomendadas incluyen un inventario de árboles actualizado regularmente para permitir un enfoque de mantenimiento proactivo, que debe definirse claramente en las ordenanzas de la ciudad junto con las medidas de protección de árboles.

Durante el desarrollo de este Plan Maestro de Bosque Urbano, el equipo se reunió con más de 350 miembros de la comunidad para comprender las oportunidades y barreras para expandir la copa de árboles urbanos. Miembros de la comunidad expresaron opiniones positivas sobre los árboles en general y el deseo de cultivar el bosque urbano, junto con preocupaciones sobre el manejo y mantenimiento adecuados, la vulnerabilidad al cambio climático y las restricciones de agua, y la aplicación de la ordenanza actual de protección de árboles.

Plan para el futuro

Después de estos análisis y en consulta con el comité directivo de la comunidad del proyecto, se identificaron dos objetivos para mejorar el programa de bosque urbano de la Ciudad. *La siguiente tabla detalla los objetivos hacia estas metas, junto con las acciones que la Ciudad puede tomar para lograr cada objetivo.*

De izquierda: Previa Alcaldesa Laura Martinez, Alcalde Ruben Abrica, Adolescentes Forestales Urbanos, y socios. Plantación de árboles de Bayshore Christian Ministries en otoño 2019. Foto: Canopy.

Objetivo 1: Cultivar un bosque urbano saludable, extenso, vibrante y diverso para proporcionar un 20% de cobertura de árboles para 2062 y un 30% para 2122.

Objetivo	Acción	Punto Clave
1. Cultivar el bosque urbano.	1. Desarrollar una estrategia de plantación de árboles en toda la ciudad para aumentar la equidad dentro de East Palo Alto y con las ciudades vecinas.	1. Plantar 200-250 árboles públicos por año
		2. Plantar 150-200 árboles privados por año
		3. Medir la cobertura arbórea de toda la ciudad y de la zona cada 10 años para realizar un seguimiento del progreso
	2. Plantar especies adaptadas al clima con la intención de plantar el árbol correcto en el lugar correcto para maximizar los beneficios para la salud pública, la vida silvestre y la resiliencia.	1. Plantar árboles apropiados para cada zona de hábitat en la ciudad
		2. Plantar la especie adecuada para el lugar de plantación específico
		3. Mantener la diversidad de especies y edades dentro del bosque urbano
	3. Integrar los árboles en el futuro diseño urbano.	1. Establecer tamaños estándar para pozos de árboles y las tiras de plantación para permitir el crecimiento saludable de los árboles
		2. Promulgar una ordenanza de sombra en estacionamientos
		3. Revisar los planes de desarrollo para la protección y plantación de árboles
2. Definir respons-abilidades y apoyar mejores prácticas de mantenimiento y protecciones para árboles públicos y privados.	4. Codificar la responsabilidad y establecer estándares para el mantenimiento y la protección proactiva de árboles públicos.	1. Codificar la autoridad y la responsabilidad de la Ciudad para el mantenimiento de árboles en viás públicas.
		2. Adoptar un enfoque proactivo de poda para el mantenimiento público de árboles
		3. Condensar el código de la Ciudad que regula árboles públicos en una sola ordenanza con todos los componentes clave, incluyendo los estándares de cuidado para árboles públicos.
	5. Revisar la ordenanza de protección de árboles y el proceso de implementación para proporcionar una fuerte protección a árboles privados.	1. Revisar la definición de árbol protegido
		2. Revisar los criterios para la eliminación de árboles protegidos
		3. Exigir que proyectos de desarrollo soliciten permisos de remoción de árboles
		4. Aumentar la capacidad del personal para administrar la ordenanza de protección de árboles mediante la contratación o contratación de un arborista certificado
		5. Revisar y aclarar las tarifas y los requisitos para las solicitudes de permisos de remoción de árboles
		6. Crear un proceso de apelaciones más sólido con el aporte de la comunidad
		7. Dar a conocer y hacer cumplir las protecciones de árboles
	6. Buscar oportunidades de financiamiento adicional.	1. Identificar oportunidades para obtener fondos adicionales para apoyar programas de bosque urbano

Objetivo 2: Conectarse con una comunidad comprometida e informada para proporcionar administración del bosque urbano.

Objetivo	Acción	Punto Clave
3. Conéctarse con la comunidad en torno a la administración de árboles.	7. Diseñar un programa de divulgación pública cohesivo e inclusivo centrado en crear conciencia sobre los beneficios de árboles y cómo y por qué los árboles son protegidos en la ciudad.	1. Participar con grupos comunitarios locales en actividades de administración de bosques urbanos
		2. Asociarse con Canopy para identificar oportunidades para expandir los programas actuales de alcance comunitarios
		3. Proporcionar información actualizada sobre la protección y el manejo de árboles en el sitio web de la Ciudad
		4. Traducir la información de árboles a idiomas que se hablan en East Palo Alto
		5. Designar un foro para que el público participe en acciones de administración y protección de árboles
		6. Identificar formas para que la Ciudad celebre árboles
	8. Convertise en Tree City USA.	1. Al igual que en la Acción 4, actualizar la ordenanza de árboles públicos para establecer estándares de cuidado
		2. Al igual que en las Acciones 5 y 7, designar o crear un foro público para revisar las solicitudes y apelaciones de eliminación de árboles, y para servir como un foro público para cuestiones relacionadas a los árboles.
		3. Al igual que en la Acción 7, planear actividades de celebración para el Día de Árboles.

East Palo Alto. Imágen: Google Earth

(top) Vice Mayor Lisa Gauthier planting her "Mayor's Tree" with arborist and former planning commissioner Uriel Hernandez at MLK Park on MLK Day 2019. (bottom) Young volunteer showing worm. Photos: Canopy.

SECTION 1:
INTRODUCTION and CONTEXT

IN THIS SECTION

Why care about the urban forest?

The urban forest is an essential part of urban infrastructure, thanks to the ability of trees to provide cost effective, natural solutions to challenges faced in urban environments (Livesley et al., 2016). By creating cooling shade, capturing air pollutants, producing fruit for people and animals, and creating beautiful places, trees are a powerful tool for building functional, attractive cities with nature.

The urban forest can serve multiple functions within a city, benefiting both nature and people. With appropriate design and management, the urban forest can mimic a natural forest and provide habitat and resources for native wildlife (Spotswood et al., 2019). Public fruit trees can increase access to local healthy foods for urban residents, and tending to these trees and sharing the harvest can help build community connections (Colinas et al., 2019). Trees can make spaces look unique and memorable, and can even boost business for stores located along tree-lined streets (Wolf, 2005). In addition, trees can improve public health by creating comfortable spaces for recreation and relaxation, creating shade under their canopies and cooling surrounding areas (Armson et al., 2012; Nesbitt et al., 2017). By providing shade, urban tree

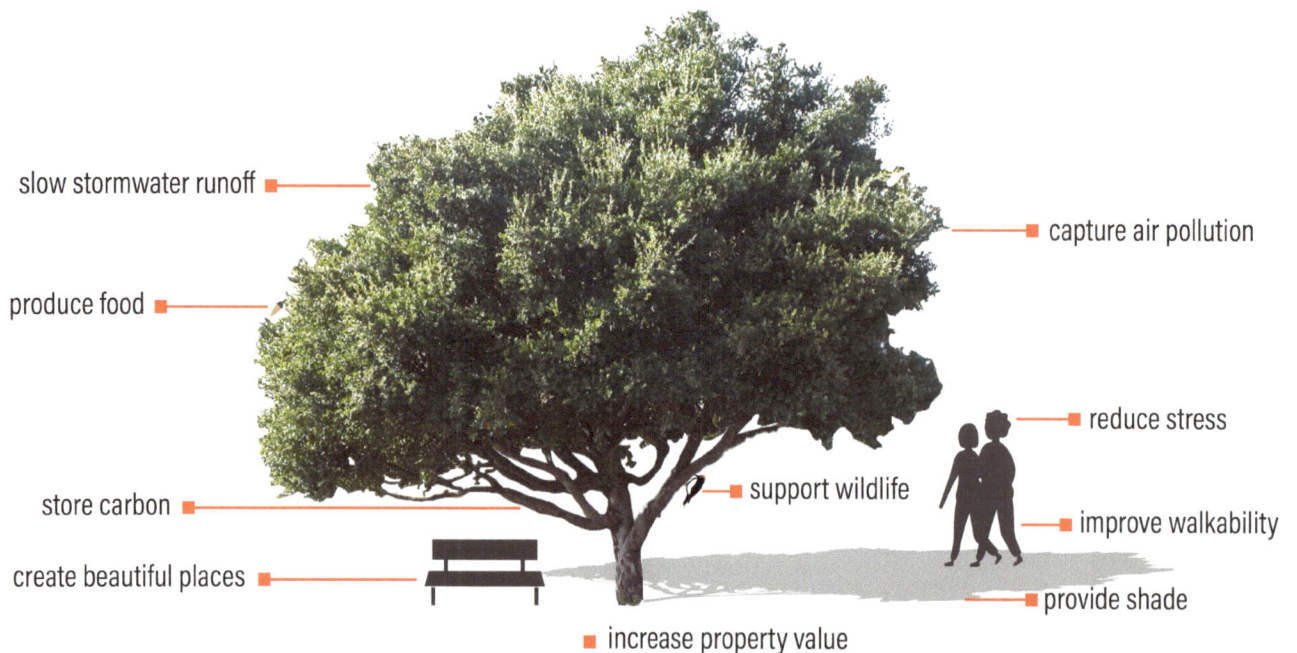

slow stormwater runoff

capture air pollution

produce food

reduce stress

store carbon

support wildlife

improve walkability

create beautiful places

provide shade

increase property value

canopy can cool buildings and people, reducing the risk of heat-related health issues such as heat stroke (Graham et al., 2016). As temperatures in the Bay Area increase due to climate change, cooling by trees can be part of a climate adaptation strategy.

The urban forest, as a component of "green infrastructure," works together with the built components of the city and can make them more effective. For example, by capturing raindrops before they reach the ground, the tree canopy slows down and reduces runoff during storms (Berland et al., 2017). Pavement on roads that are shaded by trees lasts longer before needing repairs (McPherson & Muchnick, 2005). Trees purify the air by removing floating particulate matter, and can also help combat climate change by storing carbon in their trunks, roots, and branches (Nowak & Crane, 2002; Escobedo et al., 2011). Trees can also cool buildings, saving energy and associated emissions needed for air conditioning (Roy et al., 2012).

Urban trees can have downsides as well, such as when conflicts occur between tree roots and sidewalks or building foundations. Tree planting can lead to increased housing prices, resulting in gentrification if protective policies are not in place (Donovan et al., 2021). Some trees produce allergenic pollen, which can be detrimental for the health of sensitive community members (Cariñanos & Casares-Porcel, 2011). Planning and policy can address these problems, protecting communities from displacement, supporting planting of the right trees for the right place, and providing a framework for choosing diverse and beneficial tree species.

With all of these benefits in mind, it's clear that the urban forest is a fundamental piece of the urban infrastructure, and should be cared for and invested in accordingly.

What is the urban forest?

The California Urban Forestry Act of 1978 defines the urban forest as "native or introduced trees and related vegetation in the urban and near-urban areas, including, but not limited to, urban watersheds, soils and related habitats, street trees, park trees, residential trees, natural riparian habitats, and trees on other private and public properties" (California Urban Forestry Act of 1978, 1978). While the main focus of urban forestry is on trees, all vegetation, including shrubs, flowering plants, and lawns, can be considered part of the urban forest.

Young oaks will shade this school sports field in East Palo Alto. Photo: Canopy.

An Urban Forest Master Plan for East Palo Alto

Growing and maintaining East Palo Alto's urban forest is a long term investment that must be designed to fit the needs and resources of the City. Decisions such as tree species selection, placement, and maintenance schedules affect the urban forest's ability to deliver environmental benefits, such as providing shade, reducing air pollution, and slowing stormwater runoff. In order to thrive, the urban forest requires investments including financial resources, supportive policy frameworks, and community stewardship (Ko et al., 2015; Vogt et al., 2015; Breger et al., 2019). In addition to the City's direct investment in the urban forest, fostering connections with the local community and equipping residents with tree care knowledge and resources is a key strategy for building the large part of the urban forest that is located on private property.

The Urban Forest Master Plan is a document that will guide urban forestry in East Palo Alto to maximize long-term climate, biodiversity and health benefits for the community and to ensure that urban forest management aligns with the City's strategic goals. This plan highlights existing needs and resources and presents recommendations to work toward a shared vision for a robust and equitable urban forest over the next 40 years. It includes an assessment of the current status of the city's urban forest, including tree canopy cover distribution and current management practices. The plan provides guidance for tree planting and management, drawing heavily from community input and scientific analysis. As part of addressing historical inequities in tree canopy distribution both within and between cities, the plan prioritizes planting to ensure that the urban forest's benefits are shared equitably by all the city's residents. It seeks to anticipate future challenges including those associated with development pressure and climate change to build an urban forest that will be sustainable, equitable, and resilient.

The Urban Forest Master Plan was funded by a grant from the California Department of Forestry and Fire Protection (Cal Fire) and developed by the City, San Francisco Estuary Institute, HortScience | Bartlett Consulting, and Canopy, in consultation with members of the community. Plan development primarily occurred in 2021, beginning with an evaluation of the current urban forest and community engagement process in the spring and summer (Fig. 1.1). The plan makes recommendations based on community input and scientific analysis of local ecology and climate. With the urgent need to adapt to ongoing and future challenges brought about by climate change, this plan aims to allow East Palo Alto's urban forest to deliver long term benefits to all residents and establish the city as a climate-resilient and vibrant landscape.

2018	Jan 2021	May-June 2021	June 2021	June-Aug 2021	Sept 2021 - Feb 2022	April 2022
Cal Fire Grant received	**Project work begins**	**City engagement**	**Tree canopy assessment**	**Community outreach**	**Plan development**	**Plan adoption**
The City was funded by Cal Fire through the Community and Urban Forestry Program to develop an Urban Forest Master Plan.	The team began gathering data and information on the current status of East Palo Alto's urban forest.	The team presented to City Council and Planning Commission; and developed a project steering committee including representatives from each of these bodies.	The evaluation of East Palo Alto's current tree canopy was completed, showing 13.5% tree canopy cover across the city. This was far below Menlo Park (27%) and Palo Alto (38%).	The team engaged with community members through an online survey, individual conversations, public tabling, and focus group conversations to gather input on East Palo Alto's urban forest needs and opportunities.	With feedback from the steering committee, City Council, and City staff, the team drafted and revised an Urban Forest Master Plan to evaluate the current urban forest and make recommendations for future management.	The plan was unanimously adopted by City Council on April 19, 2022.

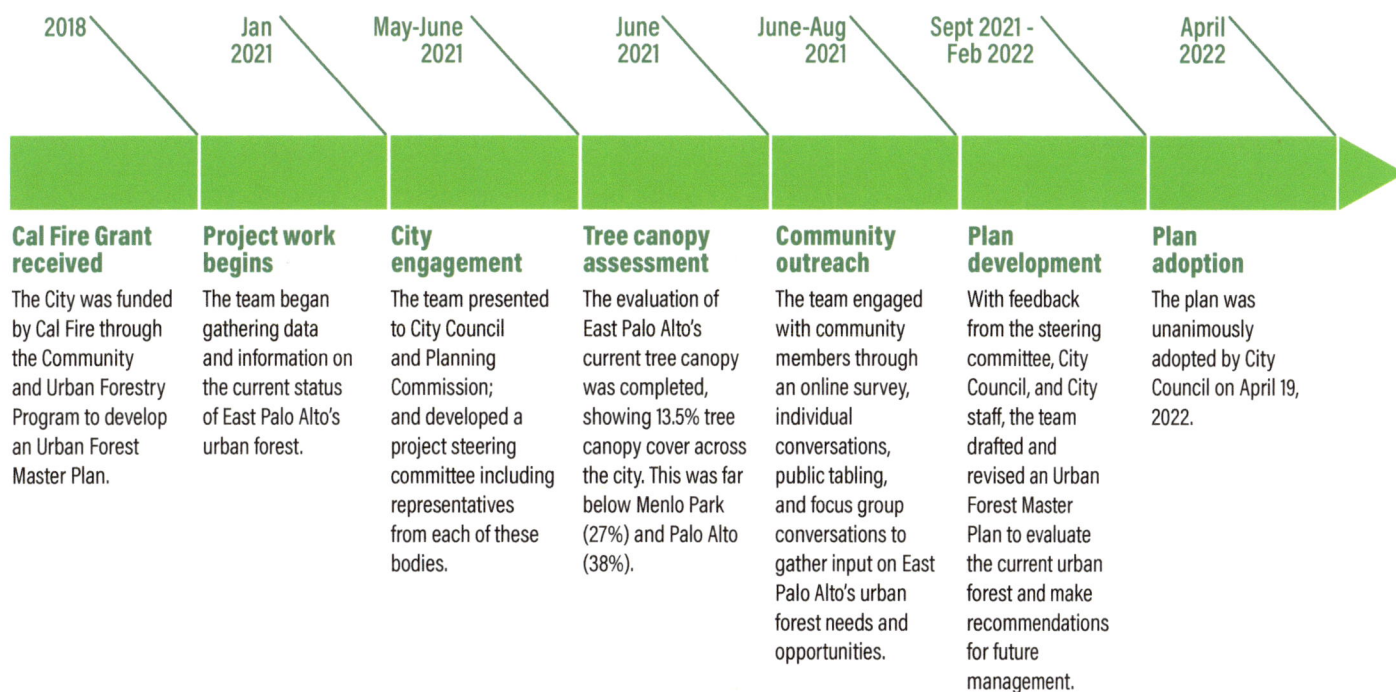

Figure 1.1. Development process for the East Palo Alto Urban Forest Master Plan.

References

Armson, D., Stringer, P., & Ennos, A. R. (2012). The effect of tree shade and grass on surface and globe temperatures in an urban area. *Urban Forestry & Urban Greening*, 11(3), 245-255. https://doi.org/10.1016/j.ufug.2012.05.002

Berland, A., Shiflett, S. A., Shuster, W. D., Garmestani, A. S., Goddard, H. C., Herrmann, D. L., & Hopton, M. E. (2017). The role of trees in urban stormwater management. *Landscape and Urban Planning*, 162, 167-177. https://doi.org/10.1016/j.landurbplan.2017.02.017

Breger, B. S., Eisenman, T. S., Kremer, M. E., Roman, L. A., Martin, D. G., & Rogan, J. (2019). Urban tree survival and stewardship in a state-managed planting initiative: A case study in Holyoke, Massachusetts. *Urban Forestry & Urban Greening*, 43, 126382. https://doi.org/10.1016/j.ufug.2019.126382

California Urban Forestry Act of 1978, Public Resources Code § 4799.06-4799.12. (1978). Retrieved August 16, 2021 from https://leginfo.legislature.ca.gov/faces/codes_displayText.xhtml?lawCode=PRC&division=4.&title=&part=2.5.&chapter=2.&article

Cariñanos, P. & Casares-Porcel, M. (2011). Urban green zones and related pollen allergy: A review. Some guidelines for designing spaces with low allergy impact. *Landscape and Urban Planning*. 101, 205–214. http://dx.doi.org/10.1016/j.landurbplan.2011.03.006

Colinas, J., Bush, P., & Manaugh, K. (2019). The socio-environmental impacts of public urban fruit trees: A Montreal case-study. *Urban Forestry & Urban Greening*, 45, 126132. https://doi.org/10.1016/j.ufug.2018.05.002

Donovan, G. H., Prestemon, J. P., Butry, D. T., Kaminski, A. R., & Monleon, V. J. (2021). The politics of urban trees: Tree planting is associated with gentrification in Portland, Oregon. *Forest Policy and Economics*, 124(August 2020), 102387. https://doi.org/10.1016/j.forpol.2020.102387

Escobedo, F. J., Kroeger, T., & Wagner, J. E., (2011). Urban forests and pollution mitigation: Analyzing ecosystem services and disservices. *Environmental Pollution*, 159, 2078-2087. https://doi.org/10.1016/j.envpol.2011.01.010

Graham, D. A., Vanos, J. K., Kenny, N. A., & Brown, R. D. (2016). The relationship between neighbourhood tree canopy cover and heat-related ambulance calls during extreme heat events in Toronto, Canada. *Urban Forestry & Urban Greening*, 20, 180-186. https://doi.org/10.1016/j.ufug.2016.08.005

Ko, Y., Lee, J.H., McPherson, E.G., & Roman, L.A., (2015). Factors affecting long-term mortality of residential shade trees: evidence from Sacramento, California. *Urban Forestry & Urban Greening*, 14, 500-507. https://doi.org/10.1016/j.ufug.2015.05.002

Livesley, S.J., McPherson, G.M., & Calfapietra, C. (2016). The urban forest and ecosystem services: Impacts on urban water, heat, and pollution cycles at the tree, street, and city scale. *Journal of Environmental Quality*, 45, 119-124. https://doi.org/10.2134/jeq2015.11.0567

McPherson, E.G., & Muchnick, J. (2005). Effects of Shade on Pavement Performance. Effects of street tree shade on asphalt concrete pavement performance. *Journal of Arboriculture*, 31(6), 303-310.

Nesbitt, L., Hwotte, N., Barron, S., Cowan, J., & Sheppard, S. R. J. (2017). The social and economic value of cultural ecosystem services provided by urban forests in North America: a review and suggestions for future research. *Urban Forestry & Urban Greening*, 25, 103-111. https://doi.org/10.1016/j.ufug.2017.05.005

Nowak, D. J., & Crane, D. E. (2002). Carbon storage and sequestration by urban trees in the USA. *Environmental Pollution*, 116: 381-389. https://doi.org/10.1016/S0269-7491(01)00214-7

Roy, S., Byrne, J., & Pickering, C. (2012). A systematic quantitative review of urban tree benefits, costs, and assessment methods across cities in different climatic zones. *Urban Forestry & Urban Greening*, 11(4), 351-363. https://doi.org/10.1016/j.ufug.2012.06.006

Spotswood, E., Grossinger, R., Hagerty, S., Bazo, M., Benjamin, M., Beller, E., Grenier, JL., & Askevold, R. A. (2019). *Making Nature's City*. SFEI Contribution No. 947. San Francisco Estuary Institute: Richmond, CA. Retrieved March 26, 2021 from https://www.sfei.org/documents/making-natures-city.

Vogt, J., Hauer, R. J., & Fischer, B. C. (2015). The costs of maintaining and not maintaining the urban forest: A review of the urban forestry and arboriculture literature. *Arboriculture and Urban Forestry*, 41(6), 293-323. https://doi.org/10.48044/jauf.2015.027

Wolf, K. L. (2005). Business district streetscapes, trees, and consumer response. Journal of Forestry, 103(8), 396-400. https://doi.org/10.1093/jof/103.8.396

2 THE CITY OF EAST PALO ALTO

History

The area that is now East Palo Alto was historically populated by the Ohlone tribe known as the Puichon. The Puichon spoke the Ramaytush dialect and lived along lower San Francisquito Creek, lower Stevens Creek, and in surrounding areas (Milliken et al., 2009). One village, Ssiputca, was located at the mouth of San Francisquito Creek, potentially near the Ravenswood area of East Palo Alto (Milliken, 1983). The Puichon and other neighboring Ramaytush Ohlone people were hunters and gatherers, moving seasonally to take advantage of a wide variety of available resources.

Prior to European colonization in the late 18th and early 19th centuries, East Palo Alto hosted diverse habitat types including oak savannas and woodlands, wet meadows and alkali meadows, willow groves, and tidal marsh (Fig. 2.1). Historically, approximately 10% of the city's area outside of the tidal Baylands was covered by willow groves. Willow groves are densely forested areas associated with high groundwater. They were dominated by arroyo willow (*Salix lasiolepis*) with a mixture of other trees and shrubs. Cooper (1926) described willow trees in the region being up to 30 feet in height, intermixed with cottonwood (*Populus* spp.), box elder (*Acer negundo*), and Oregon ash (*Fraxinus latifolia*), and underlaid with an often dense understory of California wild rose (*Rosa californica*), blackberry (*Rubus ursinus*), and ninebark (*Physocarpus capitatus*; Beller et al., 2010). In East Palo Alto, these densely forested groves were found in low-lying areas along the shorelines, where the water table was high. For example, the modern Gardens neighborhood included a large 140-acre willow grove.

In addition to willow groves hugging the shoreline, wet meadows and potentially also alkali meadows accounted for approximately 50% of the city's non-tidal area historically. Wet meadows are seasonal wetlands, with soils that remain saturated throughout the wet season—in the case of East Palo Alto, due to its fine soils and high water table. These wet meadows were characteristic of the region's black adobe soils, which were described as "extremely sticky when wet" (Lapham, 1903). Alkali meadows are also seasonally inundated wetlands, but they are associated with soils that have accumulated high levels of alkali salts, fostering a more highly-specialized community of plants that are specifically adapted to tolerate alkaline environments. While East Palo Alto's wet meadows were mostly composed of low-lying herbaceous vegetation, a "scattering growth of native oaks sometimes occurs along borders adjoining areas of lighter soil" (Lapham, 1903).

The remaining nearly 40% of East Palo Alto's historical landscape consisted of oak woodlands and oak savanna, where trees dotted a grassy landscape. These oak lands were dominated by coast live oak (*Quercus agrifolia*) and benefited from the higher ground and more well-drained alluvial deposits from San Francisquito Creek. The coast live oaks were interspersed with valley oaks (*Q. lobata*) as well, and particularly along San Francisquito Creek, the oaks were mixed with other riparian tree species, such as California box elder (*Acer negundo* ssp. *californicum*). Today, old oaks can still be found in areas of former oak woodland and savanna, particularly along San Francisquito Creek and in the Palo Alto Park neighborhood.

Colonization and increased agricultural use has greatly changed the landscape of East Palo Alto (Michelson & Solomonson, 1994). Early-arriving Spanish rancheros cleared the landscape to raise cattle, and by the mid-1800s the agricultural

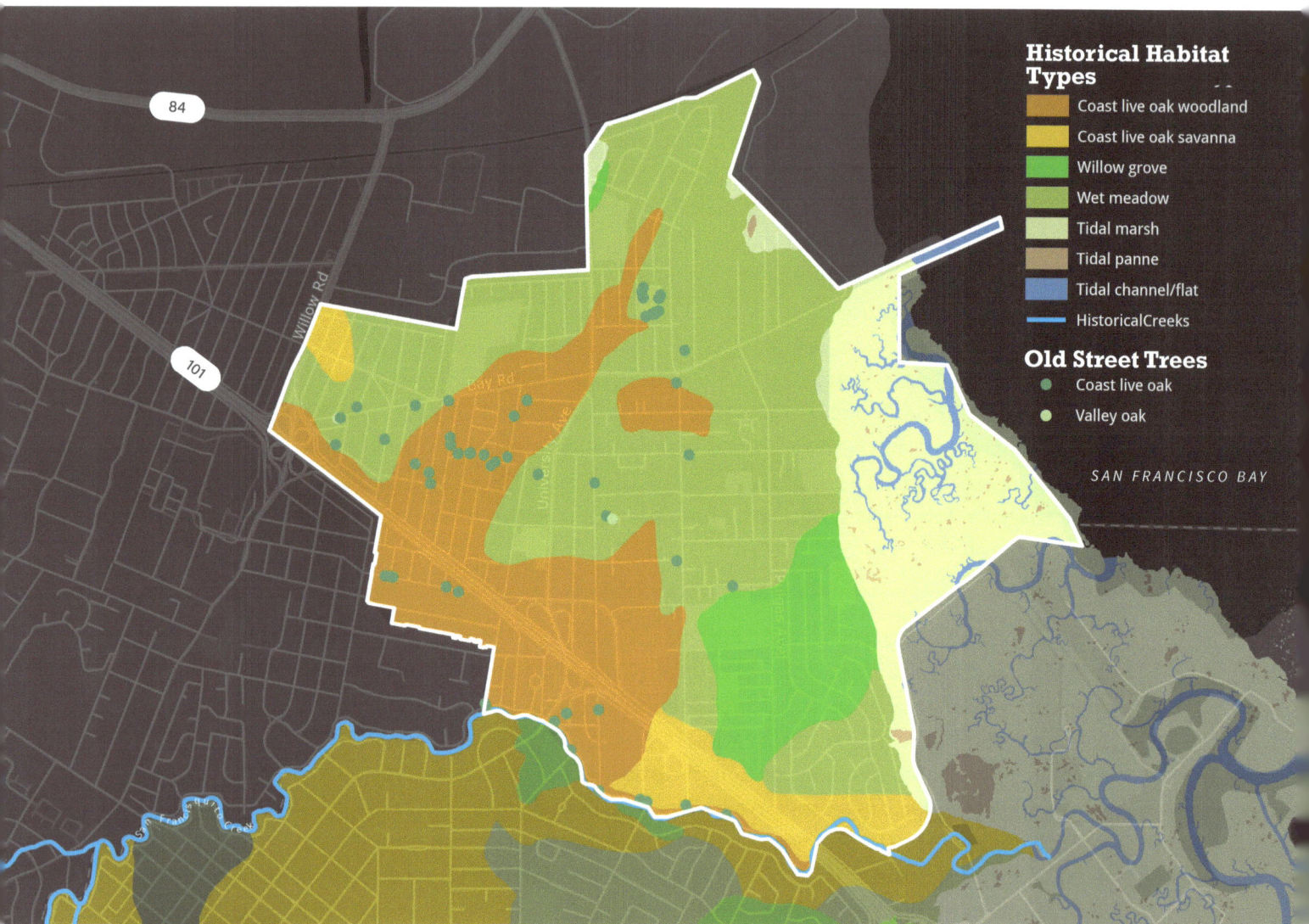

Historical Habitat Types

- Coast live oak woodland
- Coast live oak savanna
- Willow grove
- Wet meadow
- Tidal marsh
- Tidal panne
- Tidal channel/flat
- HistoricalCreeks

Old Street Trees

- Coast live oak
- Valley oak

SAN FRANCISCO BAY

0.5 miles

Figure 2.1. Map representing average ecological conditions in East Palo Alto during the early 19th century, prior to major land modifications resulting from European and American colonization and settlement. Historical ecology mapping south of San Francisquito Creek completed by Hermstad et al. (2009).

community of "Ravenswood" had developed. Willows indicated a higher water table and good quality soil for crops, and were therefore largely removed as land was converted to agricultural uses. Meanwhile, oaks were more highly valued and sometimes persisted as East Palo Alto developed around them. Over time, rural estate owners gave way to subsistence farmers, and narrow, one-acre lots were established which later facilitated the transition to a suburb. The area remained predominantly agricultural through the rest of the 19th century, with an eventual shift to floriculture.

The city's evolution into a residential suburban community began in earnest in the 1950s. As the area urbanized, University Avenue became the primary commercial corridor and the north end of town was increasingly industrialized. US 101 Highway was built in 1932 and was widened in the 1960s, isolating the city from the rest of the region both physically and culturally. After World War II, large-scale developers carved East Palo Alto into smaller subdivisions that were affordable for lower-income residents. While East Palo Alto's residential community was almost entirely white through the mid-1950s, discriminatory housing policies and practices by the federal and state real estate agencies and associations, as well as unimpeded blockbusting (real estate profiteering through stoking racial fear, causing dramatic shifts in neighborhood racial composition), led to the population's rapid transformation, such that the population was majority Black by the 1960s (Rothstein, 2017).

East Palo Alto's unincorporated status had major long-term implications for its economic and political future. As neighboring incorporated cities benefited from development capital and infrastructure improvements, East Palo Alto was excluded from the county's fiscal investments and lacked the funding to improve community spaces. The cities of Palo Alto and Menlo Park annexed large tracts of neighboring land, reducing the overall size and tax base of the City.

After decades of community advocacy and support, the City of East Palo Alto was formally incorporated in 1983. Nevertheless, the legacy of its slow development and late cityhood remain. Notably, the City's water allocations, calculated based on past usage and former contracts, were originally set in 1984, not accounting for the new city's urgent need for development, growth, and revitalization. Its critically low water allocations in the decades since have constrained East Palo Alto's ability to meet rising demand, and in 2016, its insufficient water allocations forced the City to implement a moratorium on development, including affordable housing developments and large development projects that would enhance the City's tax base.

With the City being one of the most under-resourced and recently incorporated in the region, its urban forest, like its other public infrastructure and amenities, received little investment and as a result is underdeveloped, particularly relative to conditions in neighboring cities.

The city today

East Palo Alto ecompasses 2.5 square miles at the southeastern edge of San Mateo County. It is bordered by Menlo Park to the west, Palo Alto to the south, and the San Francisco Bay to the east. The eastern edge of the city contains the natural Faber Marsh and Laumeister Marsh areas (collectively, Baylands). The city also maintains five public parks, with about 0.85 acres of public park space available per 1,000 residents. By comparison, the City of Palo Alto provides 2.6 acres of urban parkland per 1,000 residents (Sheyner 2017), and Menlo Park provides 6.48 acres per 1,000 residents (City of Menlo Park et al., 2019). East Palo Alto is separated from Palo Alto by San Francisquito Creek, which runs along the south edge of the city. Land within the city is largely developed, with almost no opportunity for new land conversion.

Figure 2.2. Map of the City of East Palo Alto, its neighborhoods, and its context in the San Francisco Bay Area region. Neighborhood names and outlines follow neighborhood designations from the Vista 2035 General Plan.

Residential area, East Palo Alto. Imagery: Google Earth.

Ecologically, the city is far different from its historical landscape. A history of agriculture and development has cleared most of the native vegetation, though areas of marsh along the Bay and some large oak trees remain. The soils too have been altered, with much of the landscape disturbed by urban development (Fig. 2.2). However, groundwater remains high, as it was when willows and wet meadows covered much of the city's low-lying, bay-adjacent land (Fig. 2.1).

The City of East Palo Alto is currently home to an estimated 30,034 residents (US Census Bureau, 2020a). 66% of the population is Hispanic/Latino, with 61% of households speaking Spanish at home. About 12% of the population is Black or African American, 5% is Asian, 5% is Native Hawaiian or other Pacific Islander, and 10% is White and not Hispanic or Latino. Median household income is low for the region at around $67,000, compared to about $160,000 for the adjacent City of Palo Alto and about $138,500 in San Mateo County (US Census Bureau, 2020b; US Census Bureau, 2020c).

The City's recent General Plan designated 12 districts or neighborhoods: primarily commercial or industrial areas, 7 primarily residential neighborhoods, and the Baylands natural preserve area (City of East Palo Alto, 2017). These neighborhoods experience different needs, challenges, and opportunities, which impact urban forest outcomes.

In the commercial areas of University Corridor, Gateway District, and 4 Corners/ Bay Road Corridor, the main opportunities for trees are along streets and in parking lots of commercial buildings and multi-family residences. Meanwhile, the Ravenswood Employment District includes large portions of vacant land, with some retail, services, and small industrial and storage facilities. Development in this area is ongoing and guided by the Ravenswood Business District/4 Corners Specific Plan, which is currently being updated (City of East Palo Alto, n.d.), creating an opportunity to integrate urban forest creation and maintenance in development planning and design. Residential areas make up most of the city's land area. Within these neighborhoods, parks, schools, and private yards provide open space for a variety of uses, including tree planting.

Figure 2.3. Map of modern soil types in East Palo Alto. Color legend below. (Kashiwagi & Hokholt, 1991; USDA NRCS 2015)

	Soil type	Depth	Drainage	Composition
	Water (#587677)	N/A	N/A	N/A
	Novato clay	Very deep	Very poorly drained	Tidal marsh soils along the margins of San Francisco Bay
	Aquic Xerorthents	Very deep	Poorly drained	Marsh soils formed in anthropogenic fill from mixed sources over bay mud from alluvium
	Orthents	Very shallow to very deep	Very poorly drained to excessively drained	Formed on uplands, including hills and ridgetops, alluvial fans, coastal terraces, flood plains, and tidal flats
	Urban land-Orthens	Very deep	Well drained	Formed on uplands, including hills and ridgetops, alluvial fans, coastal terraces, flood plains, and tidal flats
	Botella-Urban land complex	Very deep	Well drained	Formed on alluvial fans, old flood plains, and stream terraces
	Urban land-Elpaloalto complex	Very deep	Well drained	Formed on alluvial fans and flood plains
	Urban land-Elder complex	Very deep	Well drained	Formed on flood plains
	Urban land	N/A	N/A	N/A

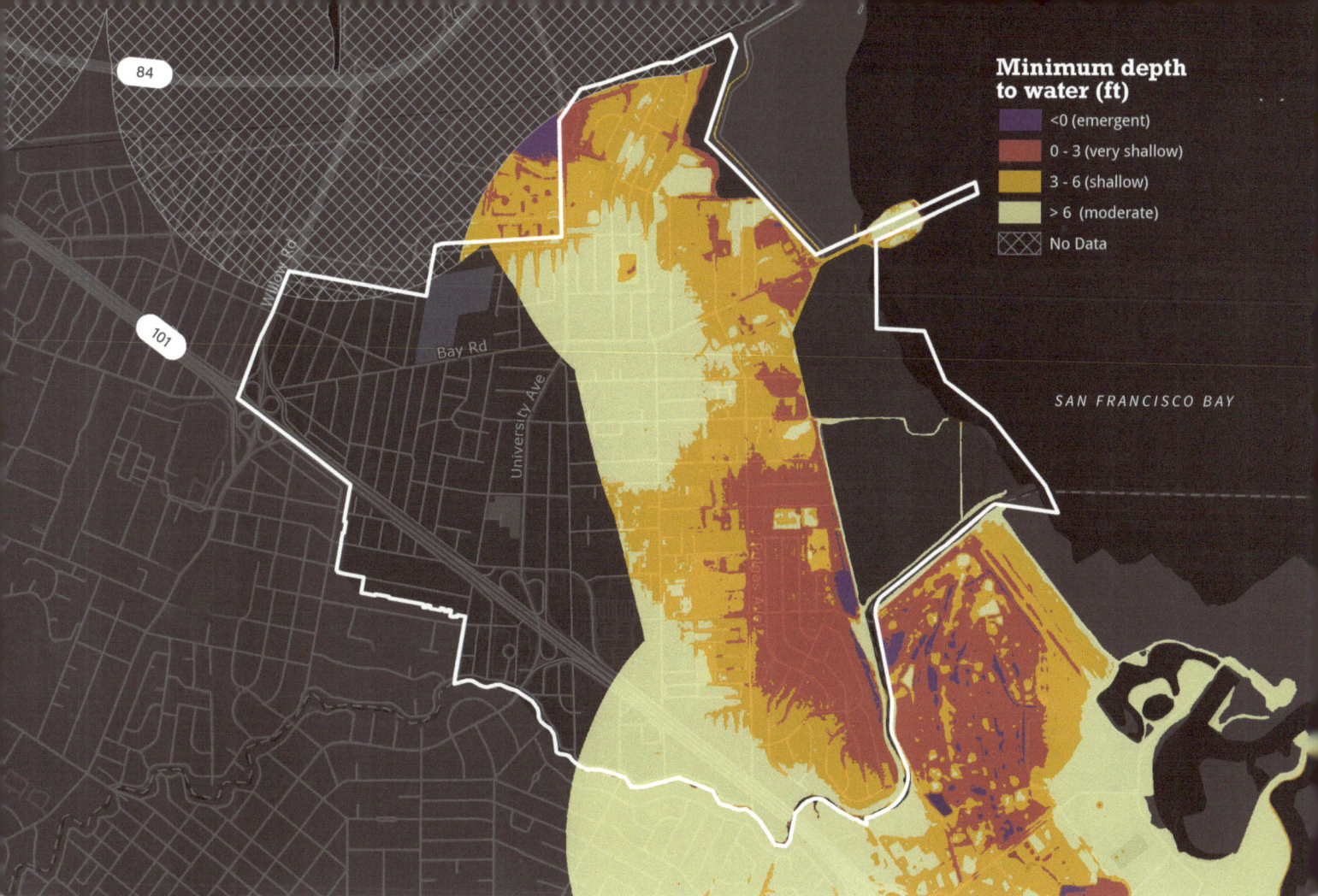

Minimum depth to water (ft)

- ■ <0 (emergent)
- ■ 0 – 3 (very shallow)
- ■ 3 – 6 (shallow)
- ■ > 6 (moderate)
- ⊠ No Data

SAN FRANCISCO BAY

0.5 miles

Figure 2.4. Map of depth to groundwater in East Palo Alto. (Plane et al., 2019)

Land adjacent to the Baylands is influenced by an additional set of constraints for tree plantings. Salt marsh harvest mouse (*Reithrodontomys raviventris*), a species protected under the Endangered Species Act, is found in the salt marshes of the San Francisco Bay. Predation by raptors and other predatory birds can pose a threat to local populations. While the significance of raptor predation on salt marsh harvest mice populations is not well understood, threats from predation are expected to intensify as sea level rise causes more intense flooding in tidal wetlands, forcing the mice to seek refuge in the uplands (USFWS, 2010). In an effort to reduce the impacts of predation, the U.S. Fish and Wildlife Service recommends that habitat-inappropriate trees, especially non-native trees, that provide perch and nest sites for raptors and corvids should be removed from marshes (USFWS, 2013).

Current and future challenges

A recent community-based vulnerability assessment highlighted risks of sea level rise, heat waves, and extreme weather events under climate change, which may affect public health and result in displacement (Thomas et al., 2020). Plans for East Palo Alto's urban forest should consider both how trees in the city will be impacted by climate change and how strategic management of the urban forest can support the city's adaptation to climate change.

East Palo Alto is exposed to severe flooding and coastal inundation, especially communities that are located near the mouth of the San Francisquito Creek. Intense rainfall, runoff from the hills, storm surge from the Bay, rising groundwater, and rising high tides are all contributing factors. As of 2015, over one third of the city's area outside of the Baylands was designated by the Federal Emergency Management Agency (FEMA) as Special Flood Hazard Area (Fig. 2.5). By 2050, sea level is most likely to rise nearly one foot relative to the

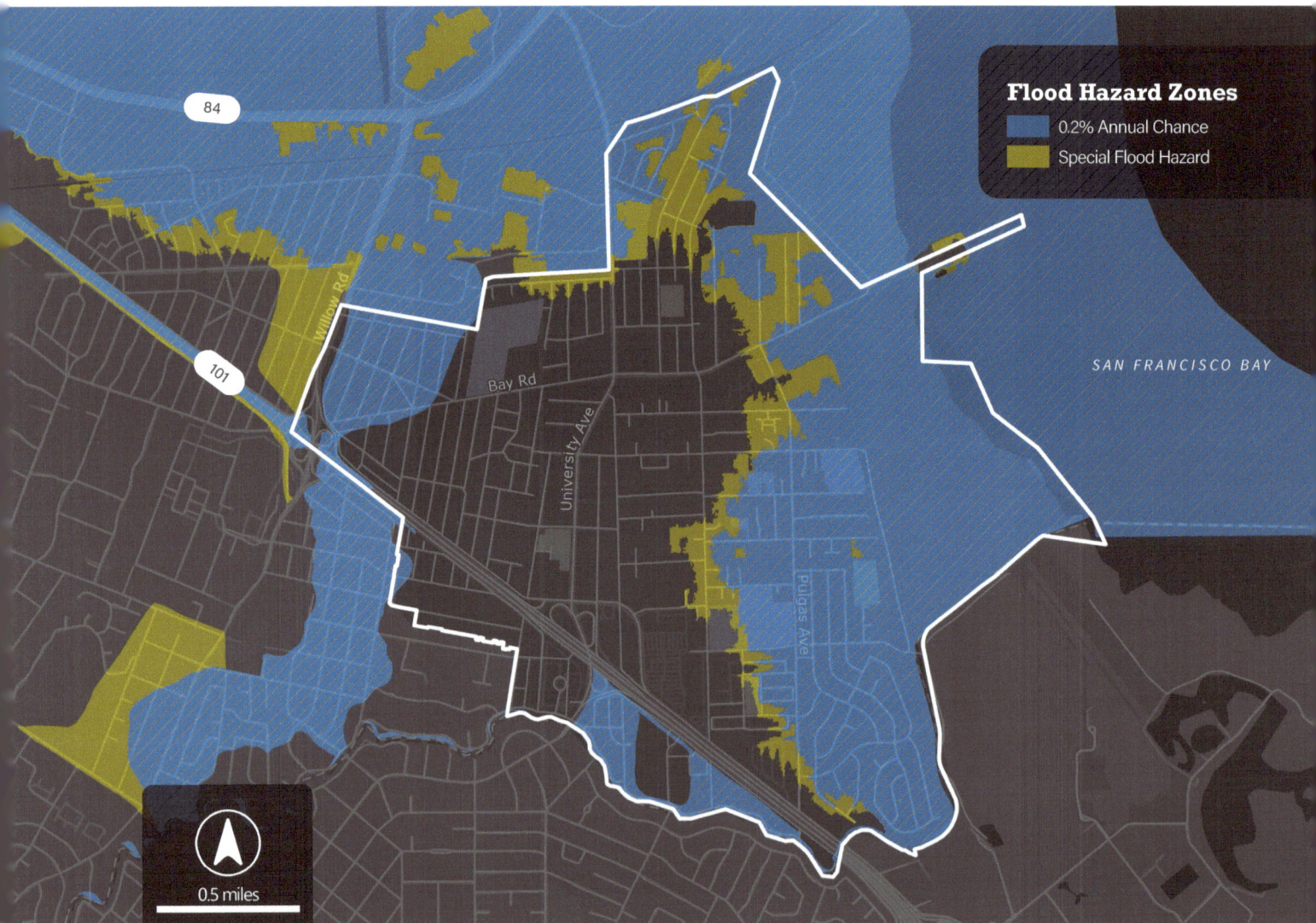

Figure 2.5. Map of flood hazard zones designated by FEMA.

year 2000, and three feet by 2100 (NRC, 2012). Nearly 60% of East Palo Alto's population is vulnerable to sea level rise, with particularly high risks of flooding and permanent inundation, storm surge, and saltwater intrusion impacting groundwater resources (California Coastal Commission, 2018; Papendick et al., 2018). The City is currently working to remedy these issues through construction of a sea wall, the first reach of which has already been constructed to provide protection to the Gardens neighborhood (Dremann, 2018).

While concerns about flooding and sea level rise imply an excess of water, East Palo Alto is also working to remedy a lack of potable water availability due to limited water rights and ongoing drought. In the recent past, East Palo Alto has struggled to procure enough water. The City's already limited water allotment from the San Francisco Public Utilities Commission has been further reduced by drought-induced restrictions. Community members have indicated that drought impacts their daily lives, such as having to comply with mandated water conservation usage (Saena, 2016). Water exchanges and transfers between East Palo Alto and other cities within and outside of the San Francisco Public Utilities Commission Regional Water System are being put to use to avoid future water shortages. In 2017, the City purchased 1 million gallons per day (MGD) from the City of Mountain View's Individual Supply Guarantee (ISG), and in 2018 the City of Palo Alto gifted 0.5 MGD of ISG to East Palo Alto under this provision. In addition to this added capacity, a Water Shortage Contingency Plan is in place to ensure sufficient future water supply (City of East Palo Alto 2020).

All water demands within the city, including tree irrigation, are currently met with potable water. Recycled water is not available to city customers, primarily because the city does not have any large parks, golf courses, or industrial uses where recycled water could have the greatest impact. Between 2014 and 2015, East Palo Alto's per capita water consumption was approximately half of the average per capita consumption among the San Francisco Regional Water System's wholesale customers (Layton & Johnson, 2019). The current per capita water use for East Palo Alto residents is approximately 60 gallons per capita per day, down from 80 gallons per capita per day prior to the 2013-2016 drought.

In addition to worsening drought, the region is projected to experience more high heat days per year in the coming decades, which may cause heat-related illnesses or worsen pre-existing conditions. While temperatures near the Bay generally remain temperate, heat waves may pose a risk in warmer parts of the city.

84

101

Willow Rd

Bay Rd

University Ave

Pulgas Ave

0.5 miles

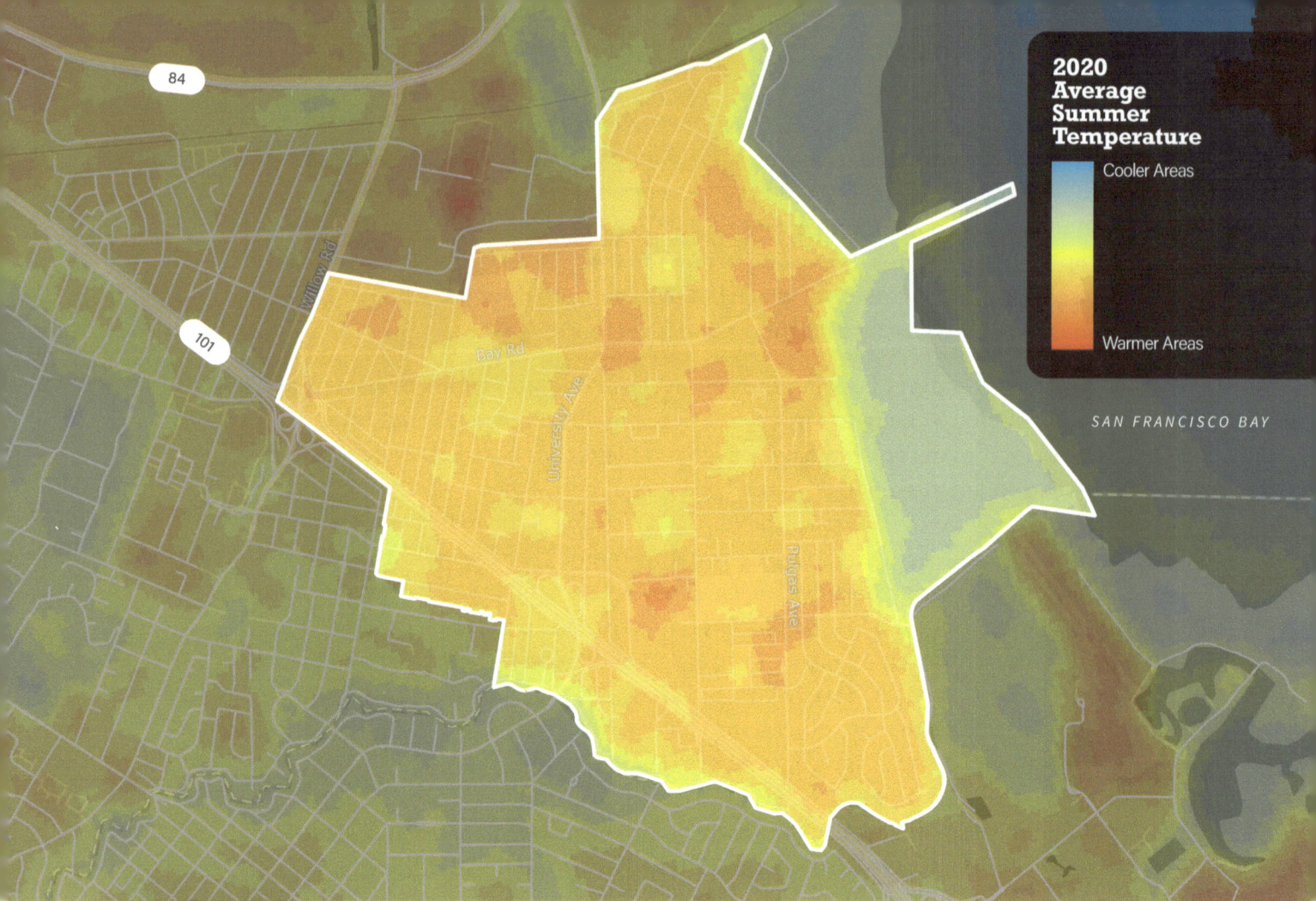

Figure 2.6. Map of 2020 summer temperatures. Values derived from Landsat 8 Provisional Surface Temperatures averaged across three summer dates: June 22, 2020, July 8, 2020, and July 24, 2020. (Landsat Level 2 Surface Temperature Science Product courtesy of the U.S. Geological Survey.)

Many East Palo Alto residents are especially vulnerable to environmental threats like flooding and extreme heat. For example, an estimated 1,860 households in East Palo Alto—or 51%—are currently considered to be financially unstable, having zero or negative discretionary income (Bick et al., 2021). As households are burdened with additional costs resulting from threats like increased flooding and need for air conditioning, households with little savings or discretionary income will be increasingly unable to absorb such costs and at risk of displacement, bankruptcy, or homelessness (Bick et al., 2021). The Centers for Disease Control and Prevention (CDC) classifies overall vulnerability of communities based on demographics recorded by the US Census Bureau, including housing and transportation characteristics, minority status and language barriers, age, disability, and socioeconomic status (CDC & ATSDR, 2020). East Palo Alto is considered highly vulnerable based on these indicators, especially relative to surrounding communities (Fig. 2.7).

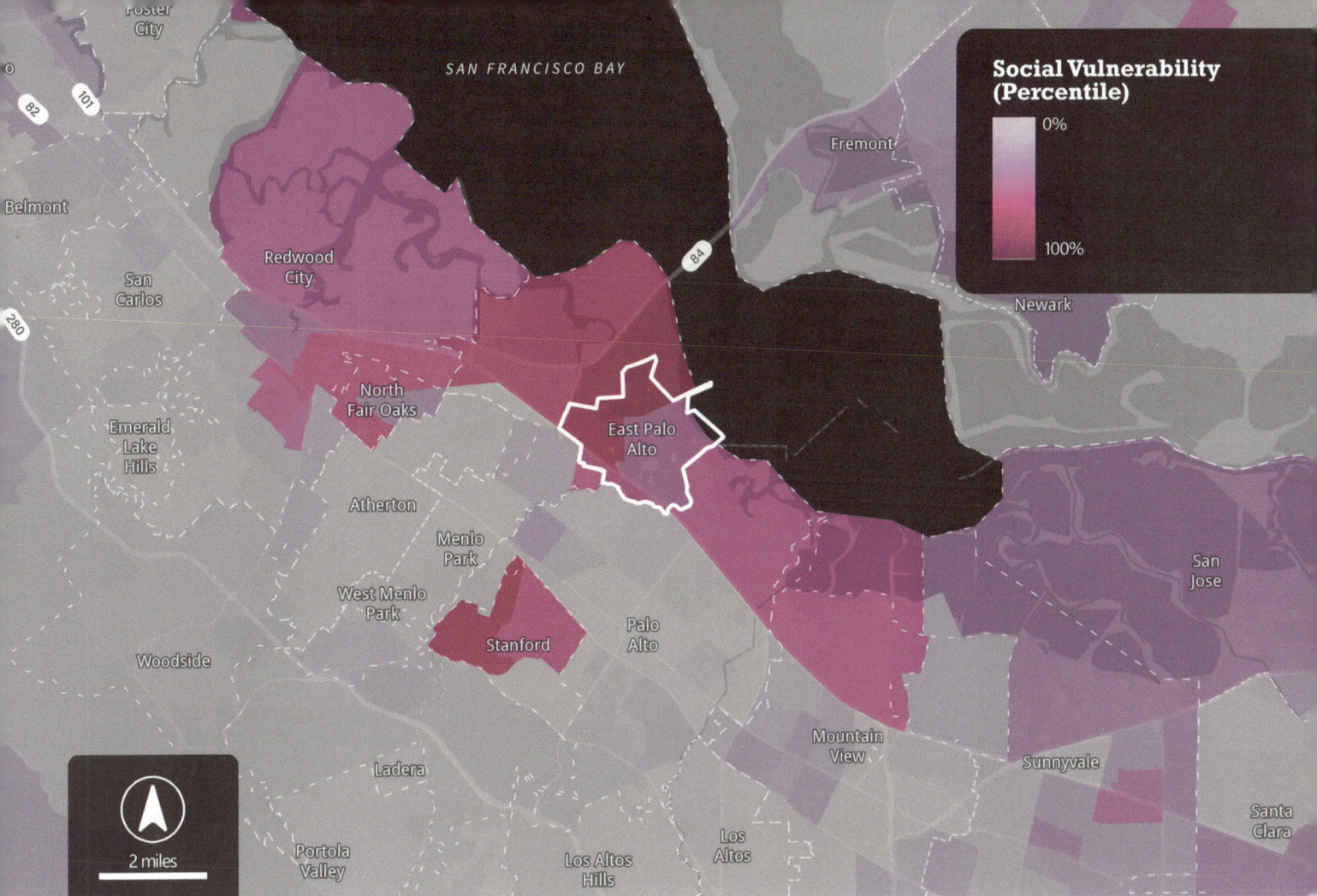

Figure 2.7. Map representing relative overall social vulnerability. Social vulnerability of each census tract is ranked based on 15 social factors reported in the US Census (CDC & ATSDR, 2020).

Strawberry tree (*Arbutus unedo*) planted on school grounds in East Palo Alto. Photo: Canopy.

Asthma (Percentile)

0% — 100%

Spatially modeled, age-adjusted rate of emergency department visits for asthma per 10,000 (2011-2013)

2 miles

Figure 2.8. Map representing relative rates of emergency department visits for asthma between 2011 and 2013 (OEHHA, 2017).

Disparities in health outcomes can also be seen between East Palo Alto and neighboring communities. Residents experience the highest rates of asthma in the county and report concerns about air pollution associated with traffic and emissions on US 101 Highway (San Mateo County, 2021a). In the survey associated with this project (see Chapter 5), 91% of respondents reported being slightly or very concerned about air pollution in the city. Over 15% of residents reported poor mental health over the previous two weeks in a 2018 survey, the highest of any city in the county, and in the highest 25% for California cities overall (San Mateo County, 2021b). Trees and other green infrastructure have potential to alleviate some of these health and wellbeing concerns in the city.

The city's current urban forest is not robust enough to meaningfully address these current and future challenges. Widespread asphalt contributes to the urban heat island effect and stormwater runoff flooding, while limited park space and tree canopy cover offer scarce areas of refuge. The city's existing urban forest provides a suite of benefits (see Chapter 3), but a strong commitment to grow the urban forest is urgently needed in order to foster a healthy, resilient city and keep pace with growing challenges.

Connections with other planning efforts

Maintenance and growth of the urban forest connects to numerous goals set by the Vista 2035 General Plan and the City's Climate Action Plan (KEMA, Inc. & City of East Palo Alto, 2011; City of East Palo Alto, 2017). These efforts also reflect county-wide goals and strategies for climate adaptation and greening. Planning for the urban forest should also influence ongoing efforts such as the City's Parks Master Plan and Ravenswood Business District/4 Corners Specific Plan Update.

The vision for East Palo Alto in the Vista 2035 General Plan describes a healthy and beautiful city with new parks, new trees and landscaping, and improved access to natural resources. Citywide greening, including the expansion of the urban forest, is a priority for achieving this vision. Due to the built-out nature of the existing infrastructure, goals focused on land use, parks, open space, and conservation are crafted to take advantage of tree planting opportunities that combine function and tangible benefits, including:

1. Enhancing the pedestrian character of streetscapes with cooling and shaded sidewalks, thereby improving the quality of life and increasing property values.

2. Creating a visual barrier while filtering particulate matter along heavy traffic thoroughfares.

3. Reducing stormwater runoff which effectively mitigates soil erosion.

4. Encouraging the planting of fruit trees in neighborhoods as a food source for residents, and foraging opportunities for wildlife, while creating the opportunity for physical connection with nature.

Guiding Principles of the Vista 2035 General Plan

Guiding Principle #8: Sustainability and environmental protection. We will strive for environmental responsibility and sustainability in our community. We are committed to preserving a healthy and ecologically flourishing planet for our children and grandchildren. We will support innovative programs and policies for environmental sustainability, climate change mitigation and adaptation, livability, and resource protection.

Guiding Principle #14: Citywide greening. We recognize the physical and mental health benefits that come from a close connection to nature, and commit to protecting and enhancing East Palo Alto's natural environment. This will include expanding the urban forest, greening public spaces, and protecting nature and habitat. We will improve our maintenance of the existing tree canopy and shift to drought-tolerant vegetation throughout City facilities.

The Parks, Open Space, and Conservation priority area includes the goal of expanding the urban forest on both public and private property, with the purpose to mitigate the impacts of climate change by increasing shaded areas at bus stops, sidewalks, public parks, and plazas. This goal includes the following policy recommendations:

- **Urban forestry.** Expand the urban forest in East Palo Alto by adding street trees and landscaping throughout the city.

- **New tree planting.** Prioritize the planting of new trees on sites designated as sensitive receptors (e.g. schools, health centers) or that are in close proximity to sources of air pollution such as freeways and heavily traveled road corridors.

- **Fruit trees.** Encourage planting of fruit trees and other edible landscaping in private development for food sources for residents and foraging opportunities for wildlife. Plant fruit trees when feasible on public property.

- **Urban forestry programs.** Support education and outreach programs to inform community members about the benefits of urban trees, including shade, improved air quality, filtration of stormwater, and wildlife habitat. Educate the community about proper tree maintenance.

Former mayor Pat Foster visits the first soundwall tree planted by the East Palo Alto Tree Initiative. Photo: Canopy.

The General Plan also identifies potential policies in other priority areas that can be achieved through strategic tree planting.

- Land Use Policy 9.9: "Tree planting. Encourage the planting and maintenance of appropriate tree species that shade the sidewalk, improve the pedestrian experience throughout the city, and enhance flood protection. Street trees should be selected that do not damage sidewalks, or block views of commercial buildings."

- Westside Policy 7.1: "Greening and streetscape. Provide additional street trees, landscaping and green space throughout the Westside to improve the area's visual appeal and increase residents' connection with nature."

- Parks, Open Space, and Conservation Policy 8.3: "Public realm shading. Strive to improve shading in public spaces such as bus stops, sidewalks and public parks, and plazas through the use of trees, shelters, awnings, gazebos, fabric shading, and other creative cooling strategies."

The City of East Palo Alto Final Climate Action Plan (KEMA, Inc. & City of East Palo Alto, 2011) shares the common goal to create a beautiful and healthy city that is sustainable by reducing greenhouse gas emissions and increasing public access to trees and the natural environment, which may have a direct or indirect effect on improving mental and physical health, increasing job productivity, and reducing crime. Tree planting is identified in the plan as a medium priority measure.

The Climate Action Plan includes the following tree-related goals:

1. Goal TL-3.1. Wider sidewalks with trees planted as a buffer between pedestrians and traffic.

2. Goal TL-4. Greater tree canopy cover through an ongoing tree planting program to reduce the heat island effect, and lower energy use to cool homes and businesses.

3. Measure TL-4.1. Regular maintenance of the urban forest to prolong the life of trees, with continual investment in the tree inventory to both facilitate removal of carbon dioxide from the air and increase long-term net carbon storage. Through photosynthesis, trees utilize the carbon to form the physical structure of the tree and return oxygen into the atmosphere. An additional perk is realized in GHG reduction credits available under the Climate Action Reserve's urban forestry protocol.

References

Beller, E. E.; Salomon, M.; Grossinger, R. M. (2010). Historical Vegetation and Drainage Patterns of Western Santa Clara Valley: A technical memorandum describing landscape ecology in Lower Peninsula, West Valley, and Guadalupe Watershed Management Areas. SFEI Contribution No. 622. San Francisco Estuary Institute: Richmond, CA. Retrieved March 26, 2021 from https://www.sfei.org/documents/historical-vegetation-and-drainage-patterns-western-santa-clara-valley-technical

Bick, I. A., Santiago Tate, A. F., Serafin, K. A., Miltenberger, A., Anyansi, I., Evans, M., Ortolano, L., Ouyang, D., Suckale, J. (2021). Rising seas, rising inequity? Communities at risk in the San Francisco Bay Area and implications for adaptation policy. *Earth's Future*, 9(7), e2020EF001963. https://doi.org/10.1029/2020EF001963

California Coastal Commission. (2018). *Interpretive Guidelines for Addressing Sea Level Rise in Local Coastal Programs and Coastal Development Permits.* California Coastal Commission. Retrieved October 26, 2021 from coastal.ca.gov/climate/slrguidance.html

[CDC & ATSDR] Centers for Disease Control and Prevention, & Agency for Toxic Substances and Disease Registry. (2020). *CDC/ATSDR social vulnerability index 2018 California.* Centers for Disease Control and Prevention & Agency for Toxic Substances and Disease Registry, Geospatial Research, Analysis, and Services Program. Retrieved August 26, 2021 from *https://www.atsdr.cdc.gov/placeandhealth/svi/index.html*

City of East Palo Alto. (2017). *Vista 2035 East Palo Alto General Plan.* City of East Palo Alto. Retrieved August 19, 2021 from https://www.ci.east-palo-alto.ca.us/econdev/page/general-plan-2035-east-palo-alto

City of East Palo Alto. (n.d.). *Ravenswood Business District / 4 Corners Specific Plan update.* City of East Palo Alto. Retrieved September 30, 2021 from https://www.cityofepa.org/planning/page/ravenswood-business-district-4-corners-specific-plan-update

City of East Palo Alto (2020). *2020 Urban Water Management Plan.* City of East Palo. Retrieved August 20, 2021 from https://www.cityofepa.org/sites/default/files/fileattachments/public_works/project/14661/final_epa_uwmp_2020_20210701_w_attachments.pdf

City of Menlo Park, Gates and Associates, Blue Point Planning. (2019). *City of Menlo Park Park & Recreation Facilities Master Plan Update.* City of Menlo Park. Retrieved December 30, 2021 from https://www.menlopark.org/ArchiveCenter/ViewFile/Item/12107

Cooper, W. S. (1926). Vegetational development upon alluvial fans in the vicinity of Palo Alto, California. *Ecology* 7(1), 1-30. https://doi.org/10.2307/1929116

Dremann, S. (2018). *Officials unveil first phase of San Francisquito Creek flood protection.* Palo Alto Weekly. Retrieved January 5, 2022 from https://paloaltoonline.com/news/2018/12/16/officials-unveil-first-phase-of-san-francisquito-creek-flood-protection

Hermstad, D., Cayce, K., & Grossinger, R. (2009). *Historical ecology of Lower San Francisquito Creek, phase 1.* Technical memorandum accompanying project GIS Data, Contribution No. 579. San Francisco Estuary Institute, Oakland, California. Retrieved February 22, 2021 from https://www.sfei.org/sites/default/files/biblio_files/San_Fran_Creek_Hist_Eco_Tech_Memo_final_web_0.pdf

KEMA, Inc., & City of East Palo Alto. (2011). *City of East Palo Alto final climate action plan: Twenty-three actions to address our changing climate.* City of East Palo Alto. Retrieved August 20, 2021 from https://www.cityofepa.org/econdev/page/climate-action-plan-adopted-2011

Kashiwagi, J. H., & Hokholt, L. A. (1991). *Soil survey of San Mateo County, eastern part, and San Francisco County, California.* United States Department of Agriculture, Soil Conservation Service. Retrieved September 1, 2021 from https://www.nrcs.usda.gov/Internet/FSE_MANUSCRIPTS/california/CA689/0/sanmateo.pdf

Layton, F. M., & Johnson, A. A. (2019). *From crisis to solutions: a case study of East Palo Alto's water supply.* Silicon Valley Community Foundation. Retrieved September 2, 2021 from https://www.siliconvalleycf.org/sites/default/files/publications/east-palo-alto-water-report-reader.pdf

Lapham, M. H. (1903). *Soil survey of the San Jose Area, California.* Washington: U.S. Department of Agriculture, Bureau of Soils.

Michelson, A., & Solomonson, K. (1994). *City of East Palo Alto historic resources inventory report.* San Mateo County Historical Association and San Mateo County Historic Resources Advisory Board. Retrieved August 13, 2021 from https://www.cityofepa.org/sites/default/files/fileattachments/community_amp_economic_development/page/2961/full_report.pdf

Milliken R. T. (1983). *The spatial organization of human population on Central California's San Francisco Peninsula at the Spanish arrival.* Sonoma State University.

Milliken R. T., Shoup, L. H., & Ortiz, B. R. (2009). *Ohlone/Costanoan Indians of the San Francisco Peninsula and their neighbors, yesterday and today.* National Park Service Golden Gate National Recreation Area.

NRC (National Research Council). (2012). *Sea-Level Rise for the Coasts of California, Oregon, and Washington: Past, Present, and Future.* Prepared by the Committee on Sea Level Rise in California, Oregon, and Washington and the National Research Council Board on Earth Sciences and Resources and Ocean Studies Board Division on Earth and Life Studies.

[OEHHA] Office of Environmental Health Hazard Assessment. (2017). CalEnviroScreen 3.0. California Environmental Protection Agency, Office Of Environmental Health Hazard Assessment. Retrieved Aug 26, 2021 from https://oehha.ca.gov/calenviroscreen/report/calenviroscreen-30.

Papendick, H., Sharma, J., Raider, C., Andrade, A., Hashizume, E., Plascencia, M., Prowitt, S., Carter, T., Malinowski, K., Ludy, J., Trahan, A., Sprague, H. (2018). *County of San Mateo sea level rise vulnerability assessment.* County of San Mateo, Office of Sustainability. Retrieved July 7, 2021 from https://seachangesmc.org/wp-content/uploads/2018/03/2018-03-05-mp-SLR_VA_Report_2.2018_v4_WEB.pdf

Plane, E., Hill, K., & May, C. (2019). A rapid assessment method to identify potential groundwater flooding hotspots as sea levels rise in coastal cities. *Water*, 11(11), 2228. https://doi.org/10.3390/w11112228

Rothstein, R. (2017). *The Color of Law: A Forgotten History of How Our Government Segregated America.* Liveright, New York.

Saena, F. V. (2016). Community vulnerability and adaptation to climate change in East Palo Alto. Masters Thesis, Duke University.

San Mateo County. (2021a). *Adults with current asthma*. San Mateo County: All Together Better. Retrieved September 30, 2021 from http://www.smcalltogetherbetter.org/indicators/index/view?indicatorId=79&localeId=153985

San Mateo County. (2021b). *Poor mental health: 14+ days*. San Mateo County: All Together Better. Retrieved September, 30 2021 from http://www.smcalltogetherbetter.org/indicators/index/view?indicatorId=1835&localeId=153985

Sheyner, G. (2017). The spreading empire. *Palo Alto Online*. Retrieved December 30, 2021 from https://www.paloaltoonline.com/news/2017/03/31/the-spreading-empire

Thomas, P., Bayuk, K., Samai, E., Harrell, B., Saena, F. V., & Clifford, T. (2020). *Acterra community-based vulnerability planning pilot project report*. Urban Permaculture Institute, Acterra, and Ecology and Environment, Inc. Retrieved September 16, 2021 from https://static1.squarespace.com/static/57d1a622d1758e0dfed089fe/t/5ed039140d56e8675511258a/1590704409246/Acterra+Pilot+Project+Report+20200422.pdf

[USDA NRCS] United States Department of Agriculture, Natural Resources Conservation Service. (2015). *Supplement to the soil survey of the Santa Clara Area, California, western part*. United States Department of Agriculture, Natural Resources Conservation Service. Retrieved July 13, 2021 from https://www.nrcs.usda.gov/Internet/FSE_MANUSCRIPTS/california/santaclaraCAwest2015/Santa-Clara-CA_West.pdf

US Census Bureau. (2020a). *East Palo Alto city, California profile: 2020 American Community Survey 1- year data*. US Census Bureau. Retrieved December 30, 2021 from https://data.census.gov/cedsci/profile?g=1600000US0620956

US Census Bureau. (2020b). *Palo Alto city, California profile: 2020 American Community Survey 1- year data*. US Census Bureau. Retrieved December 30, 2021 from https://data.census.gov/cedsci/profile?g=1600000US0655282

US Census Bureau. (2020c). San Mateo County, California profile: 2020 American Community Survey 1-year data. US Census Bureau. Retrieved December 30, 2021 from https://data.census.gov/cedsci/profile?g=0500000US06081

[USFWS] U.S. Fish and Wildlife Service. (2010). *Salt marsh harvest mouse (*Reithrodontomys raviventris*) 5-year review: summary and evaluation*. U.S. Fish and Wildlife Service, Sacramento, California. Retrieved August 11, 2021 from https://esadocs.defenders-cci.org/ESAdocs/five_year_review/doc3221.pdf

[USFWS] U.S. Fish and Wildlife Service. (2013). *Recovery plan for tidal marsh ecosystems of northern and central California*. U.S. Fish and Wildlife Service, Sacramento, California. Retrieved August 11, 2021 from https://fws.gov/sfbaydelta/documents/tidal_marsh_recovery_plan_v1.pdf

(above) Brentwood Elementary students happy to plant trees. Photo: Canopy. (below) Never too young to care for trees, at East Bayshore sound wall. Photos: Canopy.

SECTION 2:
CURRENT STATUS

IN THIS SECTION

3 URBAN FOREST ASSESSMENT

Background

To understand the current state of East Palo Alto's urban forest, we used two approaches: an evaluation of the most recent street tree inventory, and an urban tree canopy assessment.

Most cities maintain an inventory of public trees, which can include the species, size, location, health, and pruning history of each tree owned and/or maintained by the city. A tree inventory is an invaluable asset for a city to be able to manage its urban forest. However, these inventories typically only include public trees, while the majority of urban trees are on private property, such as residential backyards.

Tree canopy cover is the amount of area covered by tree leaves, branches, and stems—the tree's canopy—when viewed from above (Fig. 3.1). Urban trees provide a wide range of benefits that create more livable and resilient cities, and tree canopy cover is closely related to benefits like the amount of shade provided and the amount of rainfall captured. Measuring tree canopy cover is valuable because it can easily provide a snapshot of how all trees, both public and private, are distributed through the city. This can help identify opportunities for expanding the services generated by the urban forest.

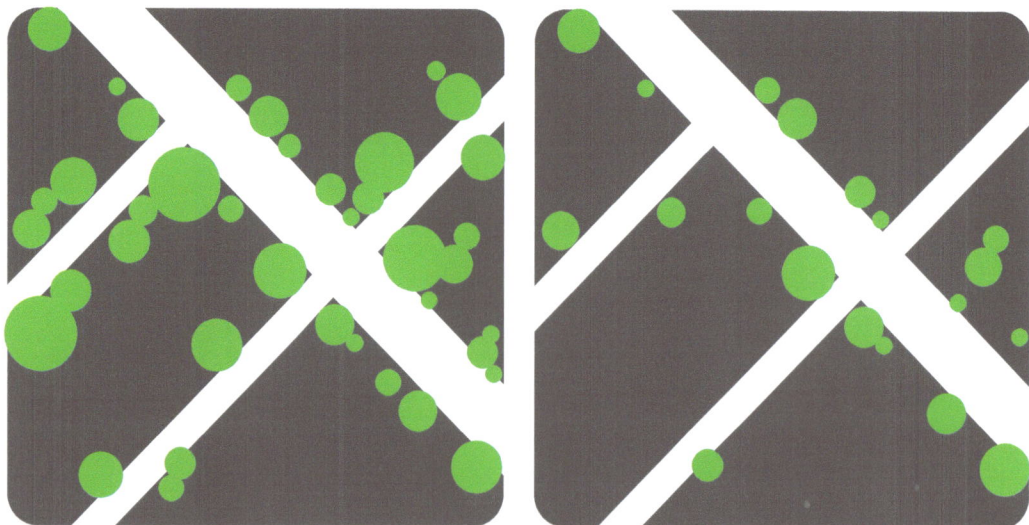

Figure 3.1. Example of a neighborhood street grid with tree canopy cover shown in green. The neighborhood on the left has a higher percentage of tree canopy than the neighborhood on the right.

In this section, we review the existing street tree inventory and evaluate the current tree canopy cover in the City of East Palo Alto to understand how the benefits of trees are distributed within the city. The following sections show the tree canopy cover of East Palo Alto compared to neighboring cities, tree canopy cover in different neighborhoods within the city, and tree canopy cover over different types of land use. We also explore how the urban tree canopy has changed since the city was incorporated to find whether tree canopy is being gained or lost over time.

Olive trees (*Olea europaea*) a few years after planting along the West Bayshore Rd sound wall (along US 101 Highway.) Photo: Canopy.

Tree canopy assessment

Measuring tree canopy cover

We summarized tree canopy cover within the different neighborhoods and land use zones of East Palo Alto. Our city boundaries were obtained from the US Census Bureau, and neighborhood and zoning boundaries came from the Vista 2035 General Plan (City of East Palo Alto, 2017). Because the Baylands are naturally unforested and situated outside of the urban landscape, the Baylands area was excluded from tree canopy cover calculations.

To measure modern tree canopy cover, we used EarthDefine's 2018 urban tree canopy cover dataset, which maps tree canopy cover in all census-defined urban areas in California (EarthDefine, 2018). This dataset maps tree canopies across the city at a high resolution (1-meter/3 feet resolution). The data are based on 2018 National Agricultural Imagery Program (NAIP) aerial imagery and Light Detection and Ranging (LiDAR) data collected by the US Geological Survey.

Tree canopy cover was calculated by adding up the total area covered by tree canopies in the city and dividing this canopy area by the total area of the city, thus representing the percent of land area covered by trees. Using EarthDefine's urban tree canopy cover dataset, we summarized tree canopy cover in East Palo Alto,

Figure 3.2. Example of 2018 EarthDefine tree urban canopy cover data. Green areas are classified as tree canopies.

excluding the Baylands. For comparison, we used similar methods to calculate tree canopy cover in Menlo Park and report Palo Alto's tree canopy cover as calculated in their 2015 Urban Forest Master Plan (City of Palo Alto, 2019).

Estimating tree canopy cover change over time

To understand how tree canopy cover in the city has changed over time, we compared aerial imagery from 1982 and 2018. The city's past tree canopy cover has not been mapped in the same way as tree canopy cover in 2018 by EarthDefine, so historical imagery from 1982 was used to estimate tree canopy around the time of the city's incorporation in 1983. We used a random point classification method to estimate canopy cover in both years (see Appendix I). We placed 750 random points on a map and determined whether each point landed on a tree or not in 1982 and 2018 (Fig. 3.3). The percentage of points that did land on a tree represents the estimated tree canopy cover.

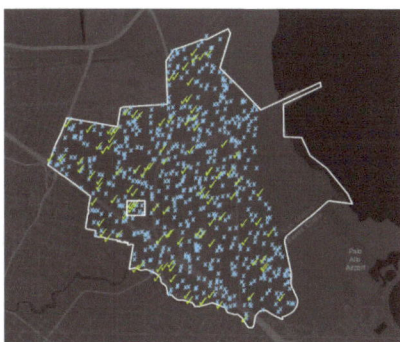

Figure 3.3. Close-up view of random sample points classified as either "tree" or "no tree" using 2018 aerial imagery. The inset map to the left shows how these points were classified across the whole city, excluding the Baylands. The same set of points was classified and used to estimate tree canopy cover in 1982 and 2018. Imagery, 2018: National Agriculture Imagery Program (NAIP), courtesy USDA Farm Service Agency.

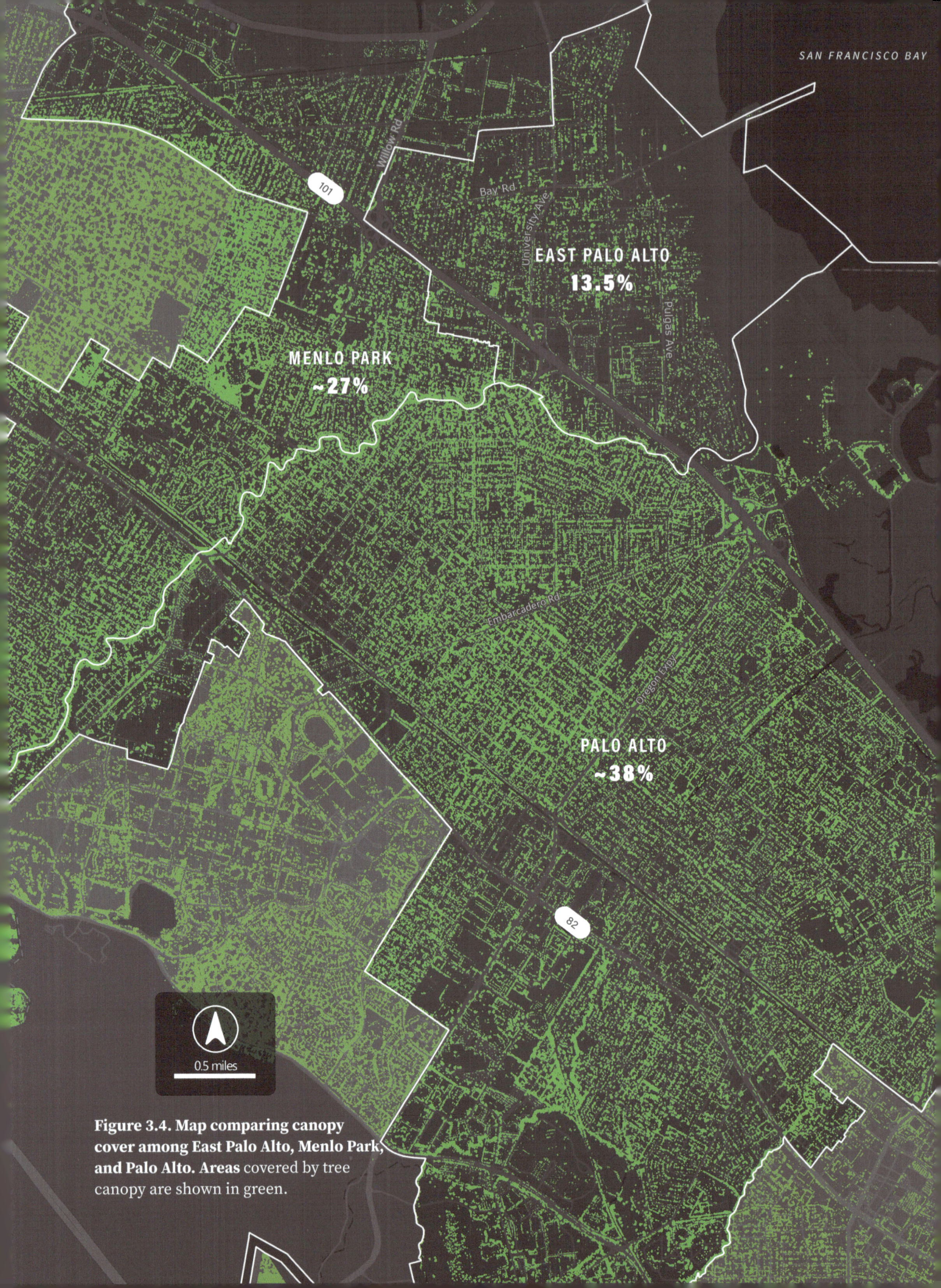

SAN FRANCISCO BAY

EAST PALO ALTO
13.5%

MENLO PARK
~27%

PALO ALTO
~38%

Willow Rd

101

Bay Rd

University Ave

Pulgas Ave

Embarcadero Rd

Oregon Expy

82

0.5 miles

Figure 3.4. Map comparing canopy cover among East Palo Alto, Menlo Park, and Palo Alto. Areas covered by tree canopy are shown in green.

Tree canopy in the city: past and present

As of 2018, East Palo Alto's urban tree canopy cover was 13.5%. In comparison, East Palo Alto's neighboring cities have much higher total tree canopy cover. To the west, Menlo Park has roughly two times more tree canopy cover than East Palo Alto (26.8%), and to the south, Palo Alto has almost three times more tree canopy cover (37.6%).

While there are many complex reasons for the discrepancy in tree cover between East Palo Alto and its neighboring cities, one important factor is the history of each city's development, incorporation, and formal and informal segregationist housing policies and practices. This history drastically impacted the allocation of resources within and between cities towards investment in community amenities, such as creating an urban forest. This legacy of disinvestment in East Palo Alto remains today and is plainly visible through its disproportionately low urban tree canopy cover.

Figure 3.5. Aerial imagery of East Palo Alto and its surroundings in 2018, visibly demonstrating the sharp contrast in greenness between East Palo Alto and its neighboring cities.

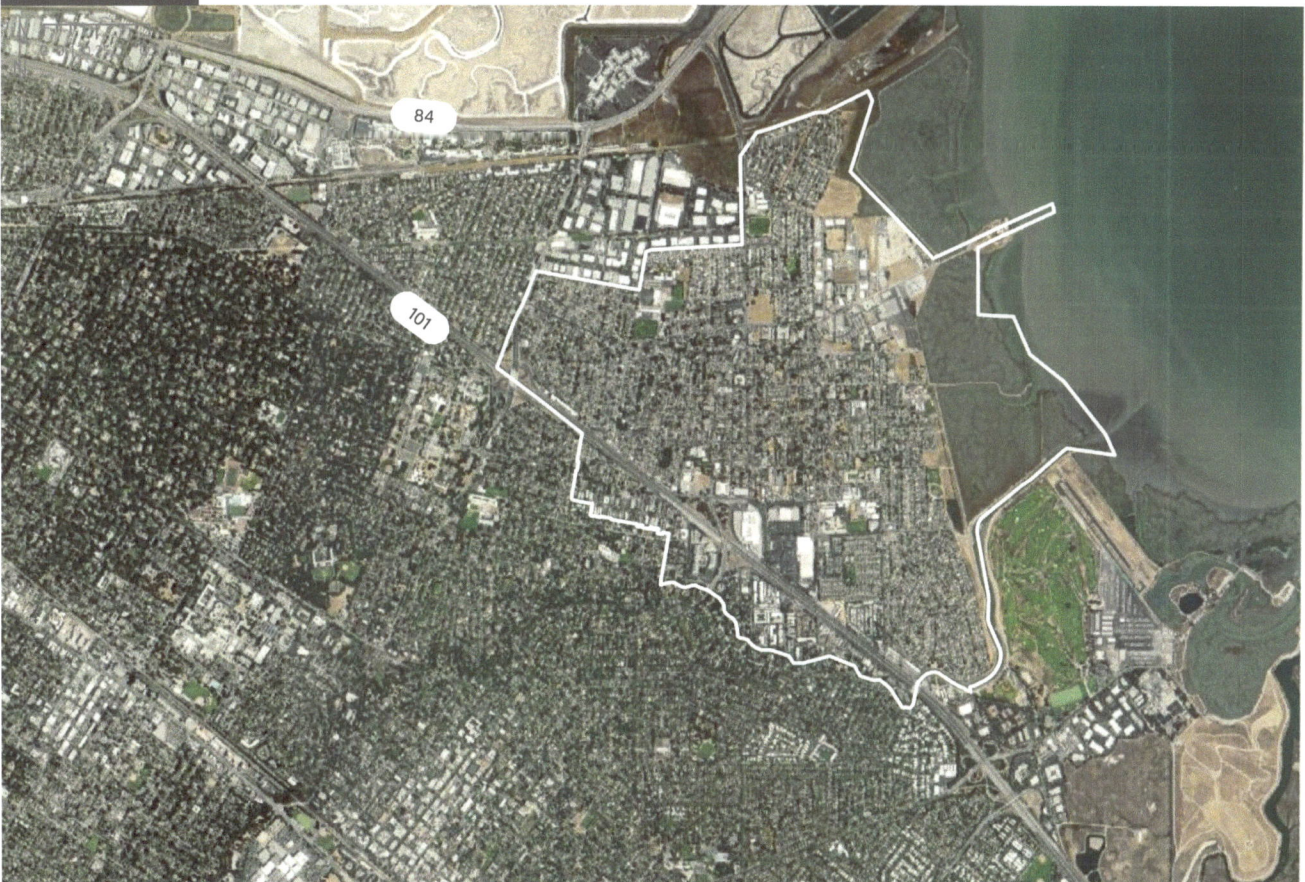

According to the *San Francisco Bay Area State of the Urban Forest Final Report* (Simpson & McPherson, 2007), Menlo Park and Palo Alto both increased tree canopy cover between 1982 and 2002, and from examining aerial imagery of the region from 1948, it is clear that these cities have been investing in their urban forests for many decades. In contrast, East Palo Alto remained largely agricultural with very little tree cover until the 1950s, when agricultural land was quickly built up with inexpensive housing. As East Palo Alto urbanized, trees were planted in moderation; however, tree canopy cover remained relatively static, with an increase of less than 5% tree canopy cover between 1982 and 2002 (Simpson & McPherson, 2007).

Extending this comparison through our tree canopy cover change analysis, we found no significant change in tree canopy cover between 1982 and 2018—a span of 36 years. Our sampling methods were only sensitive enough to detect tree canopy cover changes greater than 5%, so there may have been some very small changes, but analysis of aerial imagery in 1982 and 2018 showed tree canopy cover remained close to 14%. While overall changes across the city were small, many individual trees were certainly planted and removed over this time, and it is likely that some streets and blocks experienced overall loss or gain of tree canopy cover.

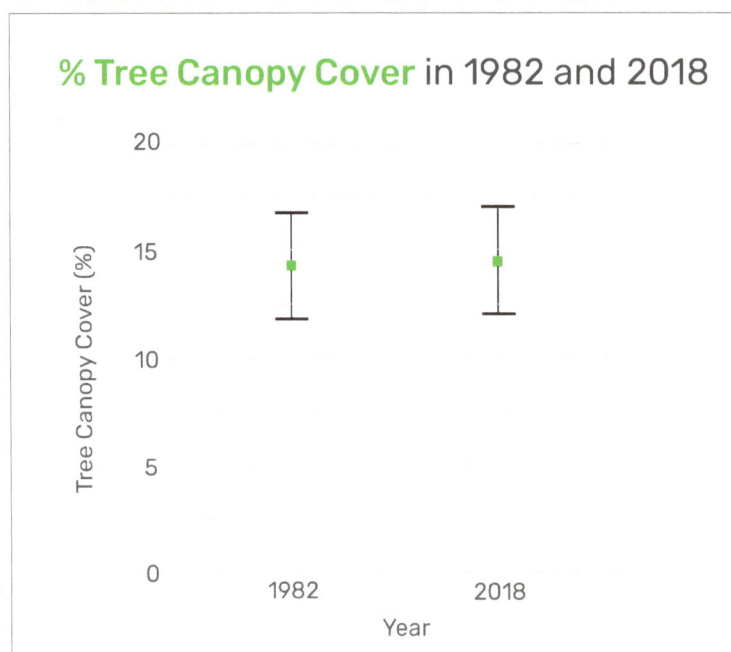

Figure 3.6. Comparison of approximate tree canopy cover in 1982 and 2018. Bars above and below the estimated canopy cover represent the range of values that we are 95% certain contains the true canopy cover.

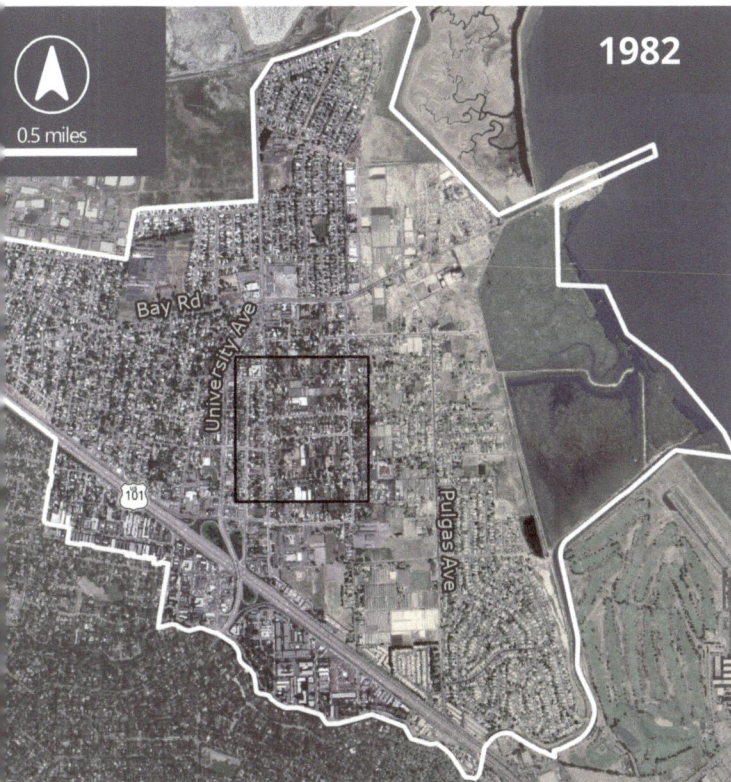

Figure 3.7. Side-by-side comparison of historical imagery from 1982 and current imagery from 2018. Imagery: 1982 imagery is courtesy Historic Aerials (historicaerials.com), and 2018 imagery is from the National Agriculture Imagery Program (NAIP), courtesy USDA Farm Service Agency.

Distribution of tree canopy cover in the city

Tree canopy cover by neighborhood

Levels of tree canopy cover differ drastically among East Palo Alto neighborhoods (as defined in the Vista 2035 General Plan; City of East Palo Alto, 2017). At the low end, the Ravenswood Employment District has only 2% tree canopy cover. The Woodland neighborhood, situated south of the US 101 Highway and directly adjacent to Palo Alto, has the highest tree canopy cover at 28%.

There is a clear pattern that the neighborhoods with the highest levels of tree canopy cover—the Woodland, Willow, and Palo Alto Park neighborhoods—are bordered by the cities of Palo Alto and Menlo Park. In comparison, neighborhoods in Palo Alto had between 13.8% (the municipal golf course) and 55.8% (Old Palo Alto) tree canopy cover in 2010, as reported in Palo Alto's 2015 Urban Forest Master Plan. It is notable that the neighborhood with the least tree canopy cover in Palo Alto has more tree canopy than East Palo Alto on average.

Figure 3.8. Map of tree canopy cover within each of East Palo Alto's neighborhoods in 2018. Areas covered by tree canopy are shown in green. The Baylands neighborhood, a naturally treeless region, was excluded from the canopy assessment.

Tree Canopy Cover by Neighborhood

Acres

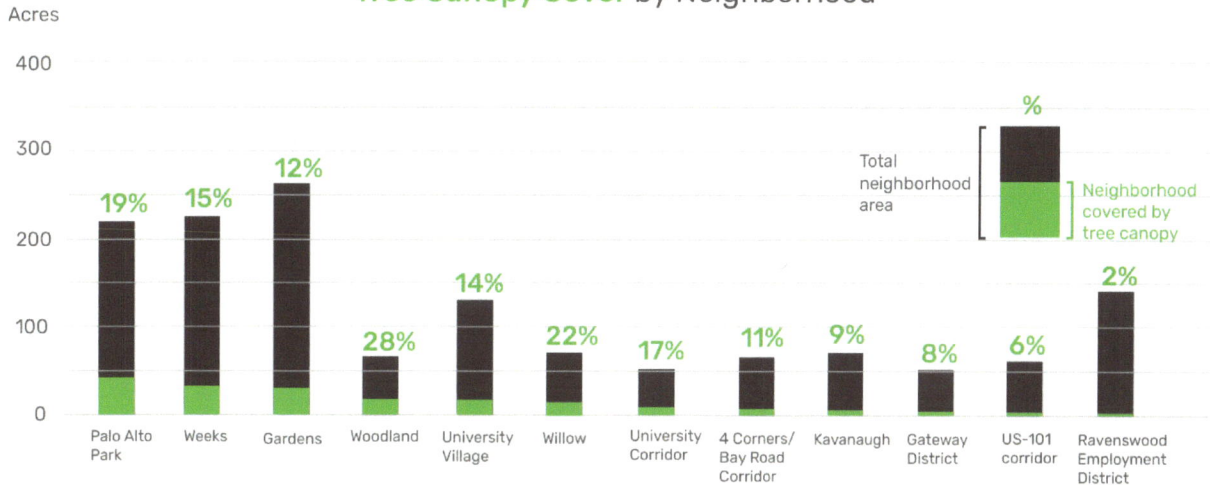

Figure 3.9. Comparison of tree canopy cover by neighborhood, listed in descending order of acres of tree canopy cover. Percentages indicate the percent of a neighborhood that is covered by tree canopy.

Tree canopy cover by zoning type

Twenty-three zoning designations delineated in East Palo Alto's Vista 2035 General Plan were grouped into eight broader zoning types for this analysis, and areas outside of zoned parcels were classified as roads and rights-of-way (Table 3.1).

About 25% of tree canopy cover in the city is on public land, while the remaining 75% is on private land. This means that private individuals are currently managing most of the trees in East Palo Alto.

Residential areas have the highest level of tree canopy cover in East Palo Alto, and the largest area of tree canopy cover is located in low density residential areas. On the other hand, the Ravenswood Business District/4 Corners Specific Plan area has the lowest level of tree canopy cover. In fact, tree canopy cover levels in all non-residential zones are lower than the city's average tree canopy cover.

Zones that are primarily owned and/or managed by public agencies tend to have low levels of tree canopy cover, highlighting an opportunity for the City to expand tree canopy cover on City-owned land. Parcels zoned for public institutional uses (e.g., public schools and government buildings) and for parks and recreation both have less than 10% tree canopy cover. Rights-of-way have a higher proportion of tree canopy cover but also cover a significant area of the city, and so have a large role to play in expanding the city's tree canopy cover.

Planting trees on parcels that are slated for redevelopment is a particularly important opportunity for expanding the urban forest in East Palo Alto. For example, the Ravenswood Business District/4 Corners Specific Plan area, which is expected to undergo significant development in the future, has only 6% tree canopy cover. During the area's redevelopment, there will be a high need to not only protect and maintain existing valuable trees, but also to capitalize on the opportunity to plant new trees.

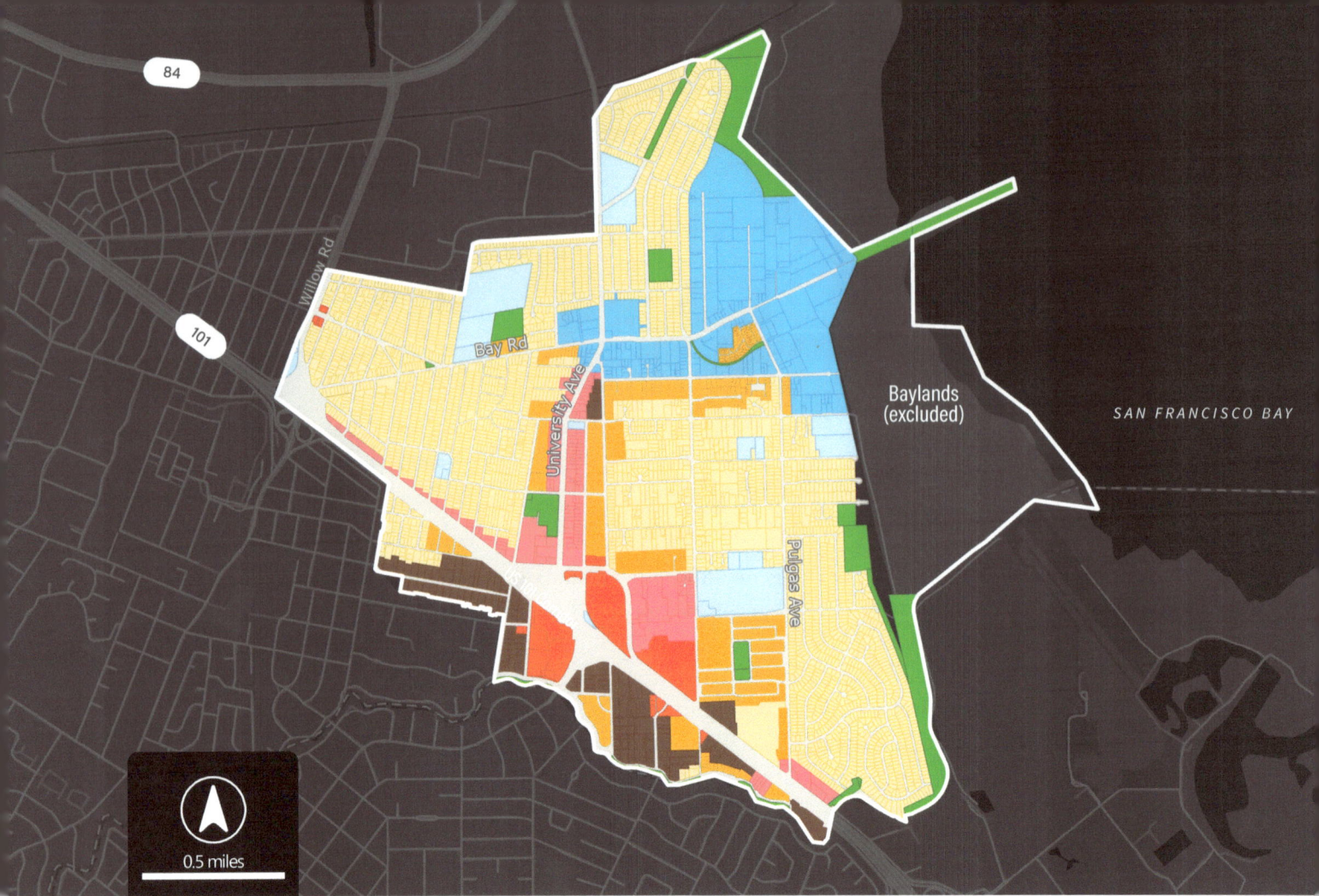

Figure 3.10. Map of the zoning types used to evaluate levels of tree canopy cover. See Table 3.1 below for classification of zoning types, and a color legend.

Table 3.1: Categorization of General Plan zoning designations into broader zoning groups for tree canopy cover measurements. Note that the Ravenswood Business District/4 Corners Specific Plan area includes commercial, industrial, mixed use, and medium density residential zones, but is grouped together as a specific plan area.

Residential, low density	Residential, medium density	Residential, high density	Commercial	Mixed use
R-LD	R-MD-1 R-MD-2 PUD	R-HD-3 R-HD-5 R-UHD	C-G C-N C-O	MUC-1 MUC-2 MUH MUL

Public institutional zone	Ravenswood Business District/4 Corners Specific Plan	Parks and recreation zone	Roads and rights-of-way
Public institutional zone	4 Corners Bay Road Central R-FO WO REC UR IT	Parks and recreation zone	Not zoned land

Tree Canopy Cover by Zone

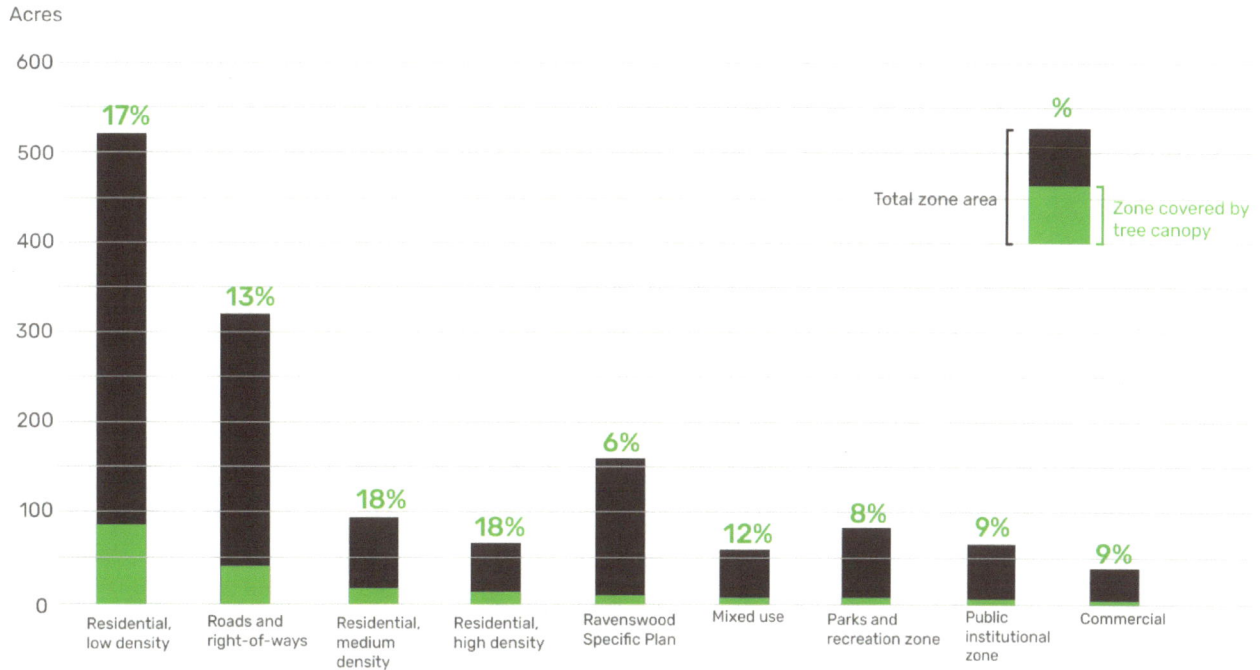

Figure 3.11. Comparison of tree canopy cover by zoning type, listed in descending order of area tree canopy cover.

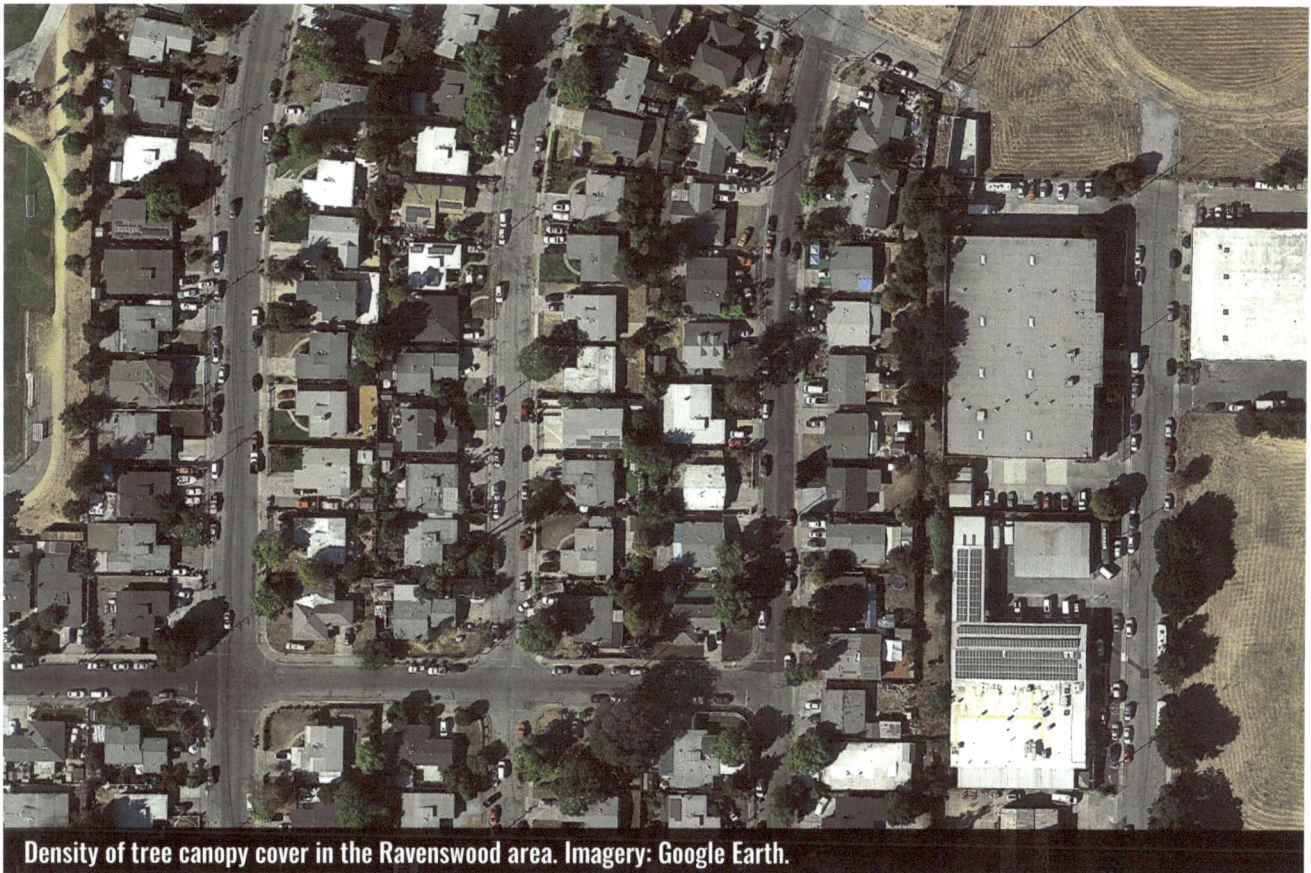

Density of tree canopy cover in the Ravenswood area. Imagery: Google Earth.

Tree canopy cover in parks and schools

East Palo Alto's parks and open spaces present important opportunities to plant trees for shade and recreation. However, the Parks and Recreation zone has relatively low tree canopy cover. Much of this area is adjacent to the Baylands, where tidal marsh habitat and species protections influence the ability to maintain high tree canopy cover (see Chapter 2). Outside of the Baylands-adjacent Cooley Landing and small pocket park, three of the four city parks (Jack Farrell Park, Bell Street Park, and Joel Davis Park) have a level of tree canopy cover that far exceeds the city average, ranging from 25% to 36%. Martin Luther King Park, which borders the Baylands, has much lower tree canopy cover, at 7%. In comparison, urban park land in Palo Alto has about 37% canopy cover overall (20% including the golf course).

Figure 3.12. Map of schools and urban parks in East Palo Alto. These spaces generally make up much of the open space in a city, and are therefore good opportunities for expanding the urban forest.

Tree Canopy Cover by Park

Acres

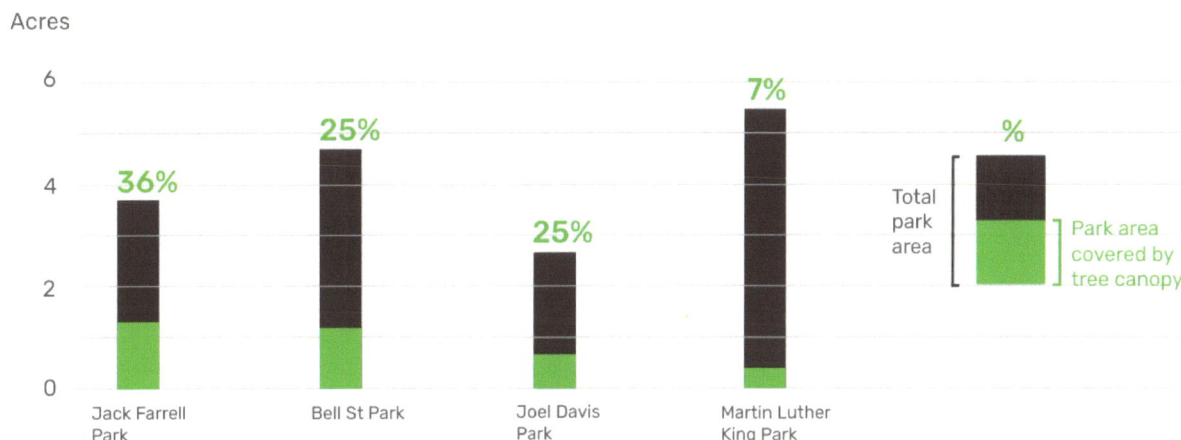

Figure 3.13. Comparison of tree canopy cover among four of East Palo Alto's parks, listed in descending order of tree canopy cover area.

While several city parks have high tree canopy cover relative to the city overall, further increasing their tree canopy would enhance their ability to act as local cooling spaces, mitigating urban heat island effects. Martin Luther King Park will require strategic tree plantings with input from ecological experts due to its location near the Baylands. Of particular interest is restoring native willow groves historically found at this site to create a natural transition, or "ecotone," between the marsh and upland areas, generating exceptional value for native wildlife. East Palo Alto is currently pursuing an open space master planning process, which could aim to increase tree canopy in city parks and could also support features such as transitional ecotones.

The Public Institutional Zone, which includes public school grounds, also has below-average tree canopy cover for the city. Tree canopy cover on school grounds is particularly important for providing healthy, shaded places for children to play, yet public schools in East Palo Alto have 7.4% tree canopy cover overall, and all schools have less than 13% tree canopy cover, the city average. For example, Cesar Chavez Ravenswood Middle School, which has the largest grounds, has only 5% tree canopy cover. In comparison, a survey of over 500 schools in Los Angeles found that schools had 11% tree canopy cover on average, and up to 34.7% tree canopy cover (Moreno et al., 2015). Only one East Palo Alto school — Oxford Day Academy, a small public charter school with a non-traditional campus — minimally surpasses this reported average, with 12% tree canopy cover. In comparison, Palo Alto school grounds have 17.5% tree canopy cover. East Palo Alto schools should be prioritized for future tree planting due to the amount of open, unbuilt area and the important benefits trees can provide for children playing outside.

Canopy Cover by School

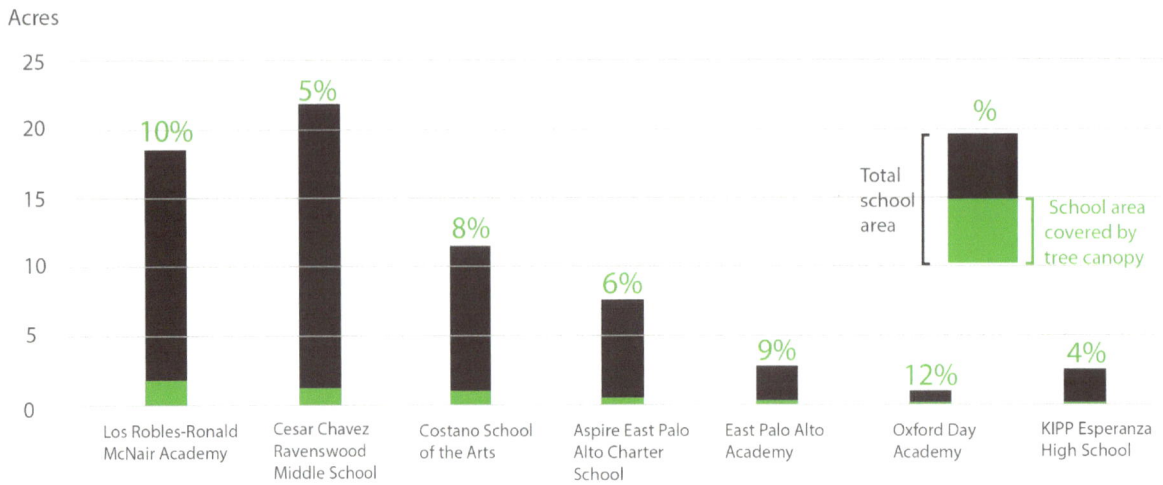

Acres

- 25
- 20
- 15
- 10
- 5
- 0

10% — Los Robles-Ronald McNair Academy
5% — Cesar Chavez Ravenswood Middle School
8% — Costano School of the Arts
6% — Aspire East Palo Alto Charter School
9% — East Palo Alto Academy
12% — Oxford Day Academy
4% — KIPP Esperanza High School

%
Total school area — School area covered by tree canopy

Figure 3.14. Comparison of tree canopy cover on school grounds within East Palo Alto, listed in descending order of tree canopy cover area.

Rain or shine, volunteer families plant trees at East Palo Alto Charter School. Photo: Canopy.

Public tree inventory

The most recent tree inventory for the City of East Palo Alto was prepared in 2013 by ArborPro, Inc., and the East Palo Alto Community Development Department, funded by Cal Fire. This was the first tree inventory conducted for the City. In the years since, the inventory has been used by Canopy to identify numerous empty planting locations and by the City to determine which trees were publicly owned. However, the inventory has not been updated to reflect changes over this time (see Chapter 4).

As part of the development of this plan, the project team collected existing records of new trees planted by Canopy and records of trees currently managed by the City's tree maintenance contractor. We also evaluated whether large trees over 30 inches in diameter recorded in 2013 still remained in 2020 using Google Street View. The inventory has been updated to reflect these changes, but a new on-the-ground evaluation of the current condition and status of public trees is needed for a full picture of the existing urban forest.

Using the current available information about East Palo Alto's public trees, we summarized the number of trees and species composition in the public domain.

Trees and acorns in protective tubes along West Bayshore sound wall, shortly after planting in 2007. Photo: Canopy.

Tree composition

The tree inventory shows that the streets and parks of East Palo Alto contain 5,745 trees, with 253 different species and cultivars.

Seventeen very common species made up 58% of the public trees, with more than 100 individuals of each species. These 17 species are common to streets and parks of the San Francisco Bay Area; none are particularly unusual or rare. They include tall, medium and small species of both deciduous and evergreen types. London plane tree (*Platanus × acerifolia*) was the most common species, making up 7% of public trees, while *Quercus* (oak) was the most common genus with 905 individuals (16% of public trees). Many different species of oaks were found in the city, including bur oak (*Quercus macrocarpa*), Canby's red oak (*Quercus canbyi*), Engelmann oak (*Quercus engelmannii*), Hungarian oak (*Quercus frainetto*), interior live oak (*Quercus wislizeni*), coast live oak (*Quercus agrifolia*), island oak (*Quercus tomentella*), netleaf oak (*Quercus rugosa*), scarlet oak (*Quercus coccinea*), Shumard oak (*Quercus shumardii*), and southern live oak (*Quercus virginiana*). Commonly used targets for promoting tree diversity suggest that the urban forest should contain no more than 10% of one species or 20% of one genus, showing that the existing public trees are sufficiently diverse. This diversity increases the resilience of the forest to future threats such as pests or disturbances that might affect closely related species.

Table 3.2. Most frequently occurring species among public trees in East Palo Alto.

Common name	Scientific name	Number of Trees
London Plane Tree	*Platanus* x *acerifolia*	427
Coast Live Oak	*Quercus agrifolia*	343
Glossy Privet	*Ligustrum lucidum*	261
Crape Myrtle	*Lagerstroemia indica*	238
Italian Cypress	*Cupressus sempervirens*	223
Catalina Cherry	*Prunus ilicifolia* ssp. *lyonii*	164
Purple-leaf Plum	*Prunus cerasifera*	163
Oak Species	*Quercus* sp.	148
Strawberry Tree	*Arbutus unedo*	138
Queen Palm	*Arecastrum romanzoffianum*	137
Water Gum	*Tristaniopsis laurina*	128
Raywood Ash	*Fraxinus angustifolia* ssp. *oxycarpa* 'Raywood'	126
Olive	*Olea europaea*	122
Coast Redwood	*Sequoia sempervirens*	107
Liquidambar	*Liquidambar styraciflua*	105
Southern Magnolia	*Magnolia grandiflora*	105
Crape Myrtle 'Muskogee'	*Lagerstroemia* x 'Muskogee'	96
All other species		2,714
Total, all trees		**5,745**

Among the remaining 236 species were several that are unique or unusual. Examples include Chinese fringe tree, Douglas-fir, European beech, Japanese tree lilac, Persian ironwood, and silver linden. Overall, 99 of the 253 species (39%) were represented by three or fewer trees, and 167 of 253 species (66%) were represented by 10 or fewer trees.

Fourteen of the 253 species are potentially native to East Palo Alto (Table 3.2). Together they comprise 565 of the 5,745 trees (10%). Coast live oak was the dominant locally native species and made up 6% of the urban forest alone. An additional 16 species are native to California, with 889 California native trees in the urban forest overall (15%).

NATIVE TREES IN EAST PALO ALTO

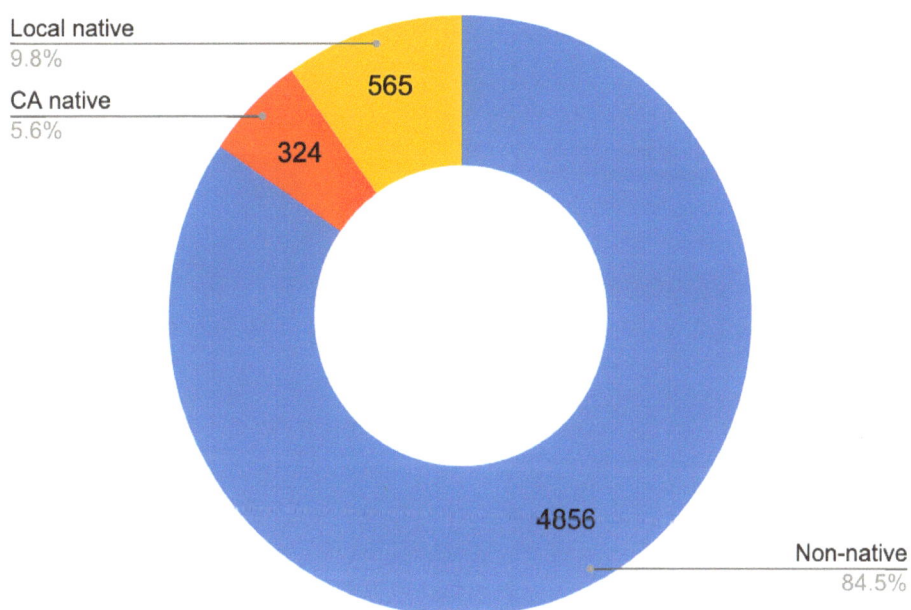

Local native
9.8%

CA native
5.6%

565

324

4856

Non-native
84.5%

Figure 3.15. Proportion of public trees in East Palo Alto that are locally native to the East Palo Alto area, native to California, or not native to California.

Some tree species found in East Palo Alto are considered potentially invasive by the California Invasive Plant Council (Cal-IPC, 2017). No trees were considered to have high invasive potential, which would have severe ecological impacts. However, six species were identified as having moderate invasive potential (edible fig (*Ficus carica*), Mexican fan palm (*Washingtonia robusta*), myoporum (*Myoporum laetum*), red clusterberry (*Cotoneaster lacteus*), tree-of-heaven (*Ailanthus altissima*), and Brazilian pepper (*Schinus terebinthifolius*)), and ten species were noted as having limited invasive potential.

Many of the trees in East Palo Alto were small in size. Just under half of trees were less than 6 inches in diameter with another 30% between 6 and 12 inches. This reflects the presence of small size trees such as crape myrtle, Catalina

Table 3.3. Tree species native to East Palo Alto found in the tree inventory.

Common name	Scientific name	Number of trees
Coast Live Oak	*Quercus agrifolia*	343
Coast Redwood	*Sequoia sempervirens*	107
California Buckeye	*Aesculus californica*	23
Valley Oak	*Quercus lobata*	23
California Bay	*Umbellularia californica*	12
Arroyo Willow	*Salix lasiolepis*	10
Bigleaf Maple	*Acer macrophyllum*	9
Box Elder	*Acer negundo*	8
Fremont Cottonwood	*Populus fremontii*	8
Willow Species	*Salix* spp.	7
Blue Elderberry	*Sambucus nigra* ssp. *cerulea*	7
Coffeeberry	*Frangula californica*	4
Oregon Ash	*Fraxinus latifolia*	3
Interior Live Oak	*Quercus wislizeni*	1

cherry, strawberry tree, and purple-leaf plum among the most frequently occurring species. These trees will remain relatively small throughout their lives. The size distribution may also reflect recent planting efforts, leading to a large number of young trees that will grow over time.

Among large trees, coast live oak was the most common, followed by coast redwood, California black walnut (*Juglans hindsii*), Canary Island date palm, southern magnolia, blue gum (*Eucalyptus globulus*), camphor (*Cinnamomum camphora*), California incense cedar (*Calocedrus decurrens*), and American sweet gum (*Liquidambar styraciflua*). Each of these species is common to landscapes of the Bay Area. There were 195 trees over 30 inches in diameter in 2013 whose presence we were able to check in 2020 via Google Street View. Of these, only 15 had been removed. Six of those 15 had been classified as being in poor health in the 2013 inventory. Meanwhile, the other 92% of large trees remain in the city today.

Various insects and diseases can threaten urban trees, potentially killing trees and reducing the overall health, value, and long-term sustainability of the urban forest. Different pests tend to target different species, meaning that the potential risk of each pest differs among cities. For East Palo Alto, 36 pests were analyzed for their potential impact and compared with pest range maps. Four of these pests are present in San Mateo County. Among the four pests, only sudden oak death (SOD) is problematic. SOD is a fungal disease with a wide host range. Coast live oak is the most important host and one of the most important species in the city. SOD is present in San Mateo County but has not been reported in East Palo Alto.

Photo 3.1. Before tree removal. Imagery: Google Street View.

Photo 3.2. After removal. Imagery: Google Street View.

Tree Planting along the US 101 Highway Corridor

A unique aspect of East Palo Alto's public trees are plantings along the US 101 Highway sound wall, located along West and East Bayshore Road on the east and west sides of the city. These plantings were part of an innovative project by Canopy, which was subsequently included in a research study on establishment success of urban trees (Roman et al., 2015). 568 trees of 17 different species were planted in 2007, with an emphasis on drought-adapted and native species. About 96% of these trees survived 6 years after planting, showing their suitability for difficult conditions along the sound wall. Irrigation and young tree care for the first few years after planting were particularly important for the great success of these trees, many of which can still be seen today.

Typical sound wall planting along US 101 Highway. Photo: Canopy.

Benefits of the urban forest

Understanding an urban forest's structure, function, and value can promote management decisions that will improve human health and environmental quality. In order to quantify some of the environmental benefits that trees provide, we used iTree Eco, which uses species information from the tree inventory along with scientific research to estimate the benefits provided by public trees. The iTree suite of urban forest analysis and management tools was developed by the US Department of Agriculture Forest Service (iTree Canopy, n.d.). iTree can be used to 1) quantify the benefits provided by trees, 2) manage and advocate for the urban forest, and 3) describe potential risks to tree and forest health.

The iTree Eco calculations are based on the species composition, size, and health of the 5,745 trees in the public tree inventory. These trees are estimated to cover about 40 acres of the city, providing about 20% of all tree canopy in the city. The remaining tree canopy cover in the city results from private trees, including those in residential yards, on school grounds, and in private parking lots. To estimate the benefits provided by the many private trees in the city, we assume that the private trees mirror the species composition, size, and health of the trees in the public tree inventory. Since there is no complete inventory of private trees, this assumption allows us to roughly estimate the important benefits trees are already providing in East Palo Alto.

The iTree analysis was constrained by the lack of cost information and limited tree data. For example, typical iTree analyses consider land use, ground and tree cover, individual tree attributes of species, stem diameter, height, crown width, crown canopy missing and dieback, and distance and direction to residential. Some of these attributes were not included in the 2013 inventory. In addition, local hourly air pollution and meteorological data were absent.

Although we cannot estimate the quantitative value of some benefits, trees in East Palo Alto undoubtedly provide a wide range of values.

BENEFITS and VALUES OF CURRENT URBAN FOREST

Compensatory value
$71,500,000

Carbon sequestration
$3,000,000

Oxygen production

Avoided stormwater runoff
$6,000

Carbon storage
$1,700,000

Air pollution removal
$95,000

Fruit production

Biodiversity support

Human health and wellbeing

Heat mitigation & building energy use

Children's academic performance

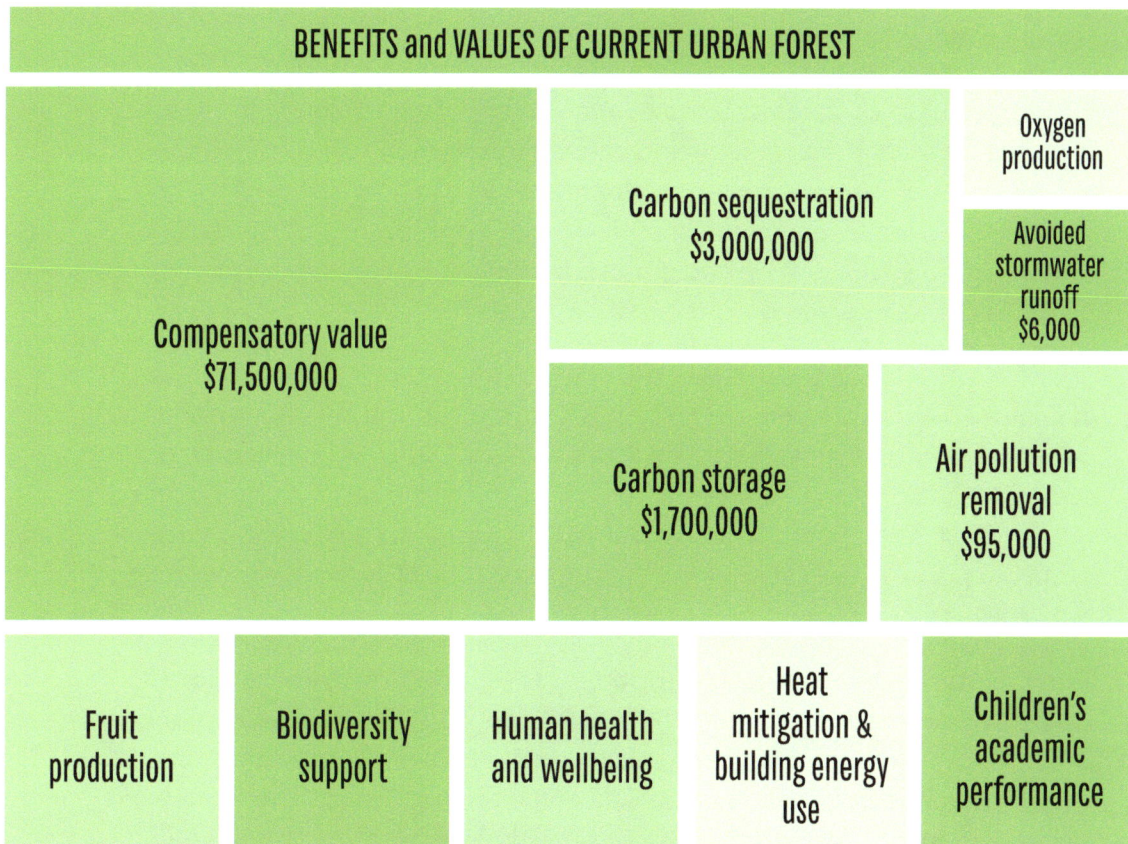

Figure 3.16. Although East Palo Alto's current urban forest is relatively small, trees in the city currently provide the community with a wide array of benefits and values, some of which can be quantified monetarily.

Urban forest structure and compensatory value

The structure of the urban forest is defined by the number of trees and their species, size, and health. Quantifying the overall value of the existing urban forest structure estimates the cost of having to replace each tree with a comparable tree, also known as the compensatory value. Larger, healthier trees have a higher value, while small, young trees are more easily and inexpensively replaced.

East Palo Alto's 5,745 public trees have an estimated compensatory value of approximately $15 million. Assuming private tree composition is the same as public tree composition, we estimate that there are about 27,500 trees total in the city, with a total value of about $71.5 million. This indicates that the average tree in the city would cost about $2,600 to replace. This high value reflects the high value of large trees remaining in the city.

Planting new trees and maintaining tree health throughout the life of each tree are both important for growing the compensatory value of the city's urban forest.

Carbon sequestration and storage

Carbon sequestration refers to how much carbon dioxide trees can capture from the atmosphere over time and convert into tree biomass through the process of photosynthesis. Carbon storage refers to the quantity of carbon stored in the tree's woody matter, including its trunk, branches, leaves, and roots. Carbon sequestration and storage are critical benefits generated by the urban forest, as they remove greenhouse gasses from the atmosphere and mitigate climate change.

In East Palo Alto, approximately 38 metric tons of carbon are sequestered by public trees each year with an associated value of $654,000, while private trees sequester an additional 140 metric tons of carbon per year. The city's entire urban forest is estimated to store over 10,000 metric tons of carbon, with a value of approximately $1.7 million. 2,140 metric tons of carbon are stored in public trees alone.

Tree maintenance is particularly important for supporting the urban forest's ability to store carbon, although there is a tradeoff in that tree maintenance activities can emit greenhouse gasses (Horn et al., 2014). In order to promote carbon storage throughout the lifecycle of a tree, the City can create systems that make use of dead and dying trees after their removal, including as long-term wood products.

Air pollution removal

East Palo Alto is burdened with poor air quality, which is linked to high community rates of asthma, among other public health and ecological issues. The urban forest can play a role in directly filtering out harmful air pollutants, in addition to cooling buildings and encouraging active transportation, which reduce energy consumption and resultant air pollutant emissions (Nowak, 2002).

Pollution removal by trees in East Palo Alto was estimated using field data and recent available pollution and weather data. The 5,745 public trees are estimated to remove 1.1 metric tons of air pollution, including ozone (O_3), carbon monoxide (CO), nitrogen dioxide (NO_2), particulate matter less than 2.5 microns ($PM_{2.5}$), and sulfur dioxide (SO_2), per year. The estimated value of air pollution removal is $20,000 per year. The entire urban forest is estimated to remove between 5 and 6 metric tons of air pollution per year, a service valued at about $95,000.

However, trees also release naturally-occurring pollutants known as volatile organic compounds (VOCs) (Nowak, 2002; Calfapietra et al., 2013). VOCs are precursor chemicals to ozone formation. In 2021, public trees in East Palo Alto will emit an estimated 1.7 metric tons of VOCs. Emissions vary among species based on species characteristics and amount of leaf biomass. Some genera such as oaks emit higher levels of VOCs.

Oxygen production

Through the process of photosynthesis, trees produce oxygen—a critical component of the air that we all need to breathe. Trees in East Palo Alto are estimated to produce about 490 metric tons of oxygen per year. However, compared to the amount of oxygen in the atmosphere overall, the production of oxygen by urban trees is not considered to be a highly valuable benefit.

Avoided stormwater runoff

Cities like East Palo Alto are typically heavily built up with impervious surfaces, such as buildings, sidewalks, and pavement, that prevent water from infiltrating into the ground. During rainfall, water runs off the surface of these impervious surfaces, creating risk of flooding and carrying contaminants into water bodies, such as rivers, lakes, and the ocean.

Trees play a role in infiltrating stormwater into the ground and slowing down its flow across the urban landscape. Trees in East Palo Alto reduce runoff by intercepting about 700,000 gallons of water per year, generating a value of about $6,000. The value of this benefit is directly tied to the annual amount of precipitation in East Palo Alto, which was 3.5 inches in 2016.

Heat mitigation

Trees help keep cities cool by providing shade and generating evaporative cooling. 30-40% tree canopy cover across the city is needed to achieve significant cooling (Ng et al., 2012; Ziter et al., 2019), though there are benefits for energy conservation, comfort, and health from even a single tree. Shade on playgrounds and in parks can protect children from heat and UV exposure (Olsen et al., 2019). Furthermore, trees also help reduce building energy consumption by shading buildings in the summer and sheltering buildings from cold winds in the winter.

The heat mitigation benefits of trees in East Palo Alto could not be quantified, but they likely reduce heating and cooling costs in buildings. They also likely reduce healthcare costs by reducing the number of heat-related illnesses and deaths (Graham et al., 2016).

Food production

Planting urban fruit trees can provide numerous benefits for cities. Fruit trees can help with food security and public health by providing a source of fresh, local produce; additionally, they give residents a stronger attachment to their city and increase their connection to nature (Colinas et al., 2019). In a more practical sense, fruit trees often take up less space than other trees, making them easier to plant on both private and public property. While their smaller size means they provide less shade, capture fewer air pollutants, and catch less rainfall, they can be a good addition to the urban forest for small spaces and where food production is especially valued. Maintenance costs, however, are the main consideration for urban fruit trees. Fruit trees can present some challenges, such as higher water use, pests, ground littering, and presence of rotting food.

The city has 366 public fruit trees. The most common species are loquat, plum, peach, and avocado (Fig. 3.17). Public fruit trees are largely located along streets, and many additional fruit trees are likely found in private yards. While this benefit could not be quantified, these trees can provide fruit to local community members if they are adequately cared for and if the fruit is harvested when ripe.

EAST PALO ALTO PUBLIC FRUIT TREES

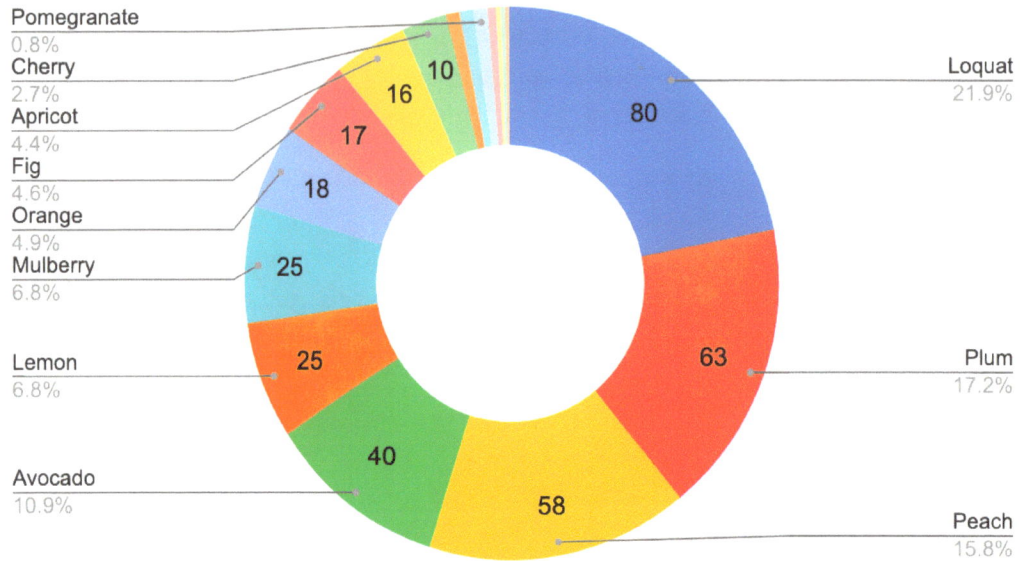

Pomegranate 0.8%
Cherry 2.7%
Apricot 4.4%
Fig 4.6%
Orange 4.9%
Mulberry 6.8%
Lemon 6.8%
Avocado 10.9%
Loquat 21.9%
Plum 17.2%
Peach 15.8%

10
16
17
18
25
25
40
58
63
80

Figure 3.17. Breakdown of East Palo Alto's 366 public fruit trees by tree species.

The urban forest also provides healthy fruit. Photo: Mbtrama, courtesy of CC 2.0.

Biodiversity support

Biodiversity is the variety of life present in a place, including both animals and plants. Urban forests can harbor relatively high levels of biodiversity (Alvey, 2006). Native urban trees in particular support insects, which in turn support predators like birds (Helden et al., 2012). The biodiversity support provided by trees increases with tree size, as large trees provide more food for wildlife, such as flowers, pollen, and nectar, and are more likely to have hollows and cavities that give animals places to nest or hide (Stagoll et al., 2012). Beyond supporting individual species, urban trees play a central role in fostering a diverse urban ecosystem that is resilient to stressors and a changing climate (Spotswood et al., 2019). Biodiverse urban landscapes are also better environments for people: studies show that people derive greater health and wellbeing benefits from biodiverse greenspaces (Sandifer et al., 2015; Wood et al., 2018).

East Palo Alto has 565 native public trees, including 367 native oaks, which are a keystone resource for a variety of fauna and flora and were historically found in this area (Grivet et al., 2007; Spotswood et al., 2017). Oaks produce acorns that are consumed by acorn woodpeckers (*Melanerpes formicivorus*), scrub jays (*Aphelocoma californica*), and California ground squirrels (*Otospermophilus beecheyi*), with cascading effects on the broader ecosystem (Tietje et al., 2005).

Although the number of animal species supported by trees in East Palo Alto is difficult to estimate, the city's large native trees, and its large oaks in particular, are likely providing food and refuge for a number of native species. According to iNaturalist, a dataset generated through community science, more than 150 native species of mammals, birds, reptiles, and amphibians and more than 200 species of native invertebrates have been recorded in East Palo Alto.

Why are large, old trees important?

Large trees have the greatest positive environmental effects per tree of all urban trees (Nowak, 2002). They act as keystone structures for supporting biodiversity in urban parks, providing food, shelter, and habitat for wildlife (Stagoll et al., 2012). In particular, cavities found in large, old trees are critical for cavity-nesting birds. Large trees remove more air pollution and sequester more carbon than small trees, create more shade, and can also intercept more rainfall, protecting water quality (Berland et al. 2017). Due to their size, they can also be particularly notable and appreciated by urban residents for their beauty and shade provision, especially in densely populated areas (Cox et al., 2019).The positive environmental effects of trees grow with the tree, which is why larger trees should be granted special protections to ensure that they can continue to provide important ecosystem services.

Health and wellbeing

Research shows that urban trees can benefit human health and wellbeing in a number of ways. First, many of the environmental benefits summarized above also lead to better health outcomes for people. Trees reduce air pollution that is associated with cardiovascular and pulmonary diseases (McDonald et al., 2016). A 10% increase in tree canopy cover can result in a 10% reduction in asthma (Ulmer et al., 2016). The urban forests' cooling effect also improves human thermal comfort and reduces the risks of heat injury (McDonald et al., 2016), especially during heatwaves (Graham et al., 2016), which are increasingly likely as global climate change accelerates.

Trees can also increase the walkability of neighborhoods and encourage greater physical activity, which leads to numerous health benefits, including reduction of obesity rates and associated health conditions (Ulmer et al., 2016, Eisenman et al., 2021). For example, research has shown that 10% more tree canopy cover increased the odds of recreational walking by 55% (Nehme et al., 2016).

Urban trees also promote mental wellbeing, including stress reduction and cognitive restoration (Wolf et al., 2020). Exposure to the urban forest has been linked with lower measures of anxiety, depression, anger, confusion, and fatigue

Figure 3.18 These images of a street in East Palo Alto were edited to show how trees can change the look and experience of a place. Photo 2 shows the addition of small trees in every other yard, photo 3 has large trees, and photo 4 has large trees in every yard. Imagery: Google Street View.

(Wolf et al., 2020), fewer mental health complaints (Akpinar et al., 2016 ; Gascon et al., 2018), and fewer prescriptions for antidepressants (Taylor et al., 2015). At the street level, studies have shown the highest stress reduction benefits between 25% and 50% tree canopy cover, far higher than currently present in East Palo Alto (Jiang et al., 2014, An et al., 2004).

While difficult to quantify, the health benefits of East Palo Alto's urban forest make it an essential component of the urban infrastructure to protect and expand.

Other benefits

Research suggests that increased tree canopy cover on school grounds can boost academic performance, leading to higher test scores, particularly for schools that serve socioeconomically disadvantaged students (Kuo et al., 2018; Sivarajah et al., 2018). In East Palo Alto, tree canopy on school grounds likely improves student performance to a modest degree, but greater tree canopy cover on school campuses could produce a more meaningful impact.

Urban trees can also increase property values and business success. Trees— either private trees located on a residential lot or public trees located in front of the home—are associated with higher home values (Wolf, 2007), as well as slightly higher monthly rent rates. In commercial areas, trees in front of businesses can give shoppers a more favorable impression of the business and increase their willingness to travel, pay for parking, and even pay higher prices for goods (Wolf, 2005). Shoppers, pedestrians, and cyclists alike prefer streets with trees along them (Lusk et al., 2020; Wolf, 2005). While these benefits can support East Palo Alto communities, increasing tree canopy cover can risk further destabilizing and displacing vulnerable populations through the process of green gentrification, compounding significant existing threats of gentrification due to current real estate and economic pressures in the region. East Palo Alto residents consistently raised concerns about the cost of housing and the need for housing security in a 2020 community vulnerability planning pilot project (Thomas et al., 2020). Certain measures (such as tenant protections and housing stabilization strategies) should be adopted prior to or in tandem with urban forest expansion to help ensure the current communities can benefit from improvements to the urban forest (Gibbons et al., 2020).

The list of benefits from the urban forest goes on—from reducing crime to extending the life of pavement on roads and parking lots (McPherson & Muchnick, 2005; Troy et al., 2012). While all of these benefits cannot be quantified and translated into dollar values, it is important to recognize the broad portfolio of values contributed to the city by its urban forest.

Conclusions

East Palo Alto's tree canopy cover is much lower than that of its neighboring cities, and these differences are rooted in the complex history of this area including the legacy of racist housing policies and practices and disinvestment in East Palo Alto. While overall tree canopy cover has not changed significantly since the city's incorporation in 1983, there is evidence of both tree removal and tree planting that suggest the urban forest has not been static. Tree canopy within the city is not equally distributed: some neighborhoods, especially those closest to Menlo Park and Palo Alto, have much higher levels of tree canopy cover.

The urban forest contains a diversity of species, but a few trees dominate. Protecting diversity in the urban forest can increase resilience to future threats like pest infestations, drought, or extreme weather. Expanding the palette of native species can help support animal biodiversity, while evaluating the outlook for different species and adjusting planting plans can improve climate resilience.

Urban trees are already providing substantial benefits in the city, and maintaining the urban forest will protect existing benefits, while increased planting can expand benefits further. An updated on-the-ground tree inventory can help assess the current health of trees in the city, which impacts their ability to provide benefits.

These findings suggest that East Palo Alto has a strong need to invest in and expand its urban forest city-wide. Because of the unequal distribution of the city's tree canopy cover, some neighborhoods and land use types will need special attention to improve their tree canopy cover levels and create more equitable access to the benefits generated by trees. Understanding the distribution of tree canopy cover in East Palo Alto historically and today can help the City identify realistic tree canopy cover targets and develop a strategic plan to expand and maintain an urban forest that supports the health and well being of the city's communities.

Volunteers at 2019 MLK Day tree planting at San Francisquito Creek. Photo: Canopy.

References

Akpinar, A., Barbosa-Leiker, C., & Brooks, K. R. (2016). Does green space matter? Exploring relationships between green space type and health indicators. *Urban Forestry & Urban Greening*, 20, 407-418.

Alvey, A. A. (2006). Promoting and preserving biodiversity in the urban forest. *Urban Forestry & Urban Greening*, 5(4), 195-201. https://doi.org/10.1016/j.ufug.2006.09.003

An, K. W., Kim, E. Il, Jeon, K. S., & Setsu, T. (2004). Effects of forest stand density on human's physiopsychological changes. *Journal of the Faculty of Agriculture, Kyushu University*, 49(2), 283–291. https://doi.org/10.5109/4588

Berland, A., Shiflett, S. A., Shuster, W. D., Garmestani, A. S., Goddard, H. C., Herrmann, D. L., & Hopton, M. E. (2017). The role of trees in urban stormwater management. *Landscape and Urban Planning*, 162, 167–177. https://doi.org/10.1016/j.landurbplan.2017.02.017

Cal-IPC. (2017). *The Cal-IPC Inventory*. California Invasive Plant Council: Berkeley, CA. Retrieved September 30, 2021 from https://www.cal-ipc.org/plants/inventory

Calfapietra, C., Fares, S., Manes, F., Morani, A., Sgrigna, G., & Loreto, F. (2013). Role of biogenic volatile organic compounds (BVOC) emitted by urban trees on ozone concentration in cities: A review. *Environmental Pollution*, 183, 71–80. https://doi.org/10.1016/j.envpol.2013.03.012

City of East Palo Alto. (2017). *Vista 2035 East Palo Alto General Plan*. City of East Palo Alto. Retrieved May 10, 2021 from https://www.ci.east-palo-alto.ca.us/econdev/page/general-plan-2035-east-palo-alto

City of Palo Alto. (2019). *Sustaining the legacy: Palo Alto Urban Forest Management Plan*. City of Palo Alto. Retrieved March 9, 2021 from https://www.cityofpaloalto.org/files/assets/public/public-works/tree-section/ufmp/urban-forest-mp-after-adoption-reduced-2-25-19-complete.pdf

Colinas, J., Bush, P., & Manaugh, K. (2019). The socio-environmental impacts of public urban fruit trees: A Montreal case-study. *Urban Forestry & Urban Greening*, 45, 126132. https://doi.org/10.1016/j.ufug.2018.05.002

Cox, D. T. C., Bennie, J., Casalegno, S., Hudson, H. L., Anderson, K., & Gaston, K. J. (2019). Skewed contributions of individual trees to indirect nature experiences. *Landscape and Urban Planning*, 185, 28–34. https://doi.org/10.1016/j.landurbplan.2019.01.008

EarthDefine. (2018). Urban Tree Canopy 2018. Earthdefine. Retrieved March 9 2021 from https://www.fs.usda.gov/detail/r5/communityforests/?cid=fseprd647385

Eisenman, T. S., Coleman, A. F., & LaBombard, G. (2021). Street Trees for Bicyclists, Pedestrians, and Vehicle Drivers: A Systematic Multimodal Review. *Urban Science*, 5(3), 56. https://doi.org/10.3390/urbansci5030056

Gascon, M., Sánchez-Benavides, G., Dadvand, P., Martínez, D., Gramunt, N., Gotsens, X., Cirach, M., Vert, C., Molinuevo, J. L., Crous-Bou, M., & Nieuwenhuijsen, M. (2018). Long-term exposure to residential green and blue spaces and anxiety and depression in adults: A cross-sectional study. *Environmental Research*, 162, 231-239. https://doi.org/10.1016/j.envres.2018.01.012

Gibbons, A., Liu, H., Malik, F., O'Grady, M., Perron, M., Palacio, E., Trinh, S., Trinidad, M. (2020). *Greening in place: Protecting communities from displacement.* Audubon Center at Debs Park, Public Counsel, SEACA-LA, and Team Friday. Retrieved May 14, 2021 from https://www.greeninginplace.com/s/GG-2020-ToolKit-FINAL.pdf

Graham, D. A., Vanos, J. K., Kenny, N. A., & Brown, R. D. (2016). The relationship between neighbourhood tree canopy cover and heat-related ambulance calls during extreme heat events in Toronto, Canada. *Urban Forestry & Urban Greening*, 20, 180–186. https://doi.org/10.1016/j.ufug.2016.08.005

Grivet, D., Sork, V. L., Westfall, R. D., & Davis, F. W. (2007). Conserving the evolutionary potential of California valley oak (*Quercus lobata* Née): A multivariate genetic approach to conservation planning. *Molecular Ecology*, *17*(1), 139-156. https://doi.org/10.1111/j.1365-294X.2007.03498.x

Helden, A. J., Stamp, G. C., & Leather, S. R. (2012). Urban biodiversity: Comparison of insect assemblages on native and non-native trees. *Urban Ecosystems*, 15(3), 611-624. https://doi.org/10.1007/s11252-012-0231-x

Horn, J., Escobedo, F. J., Hinkle, R., Hostetler, M., & Timilsina, N. (2014). The role of composition, invasives, and maintenance emissions on urban forest carbon stocks. *Environmental Management*, 55(2), 431-442. https://doi.org/10.1007/s00267-014-0400-1

i-Tree Canopy. i-Tree Software Suite v6. (n.d.). Retrieved August 5, 2021 from http://www.itreetools.org

Jiang, B., Chang, C. Y., & Sullivan, W. C. (2014). A dose of nature: Tree cover, stress reduction, and gender differences. *Landscape and Urban Planning*, 132, 26-36. https://doi.org/10.1016/j.landurbplan.2014.08.005

Kuo, M., Browning, M. H. E. M., Sachdeva, S., Lee, K., & Westphal, L. (2018). Might school performance grow on trees? Examining the link between "greenness" and academic achievement in urban, high-poverty schools. *Frontiers in Psychology*, 9, 1669. https://doi.org/10.3389/fpsyg.2018.01669

Lusk, A. C., da Silva Filho, D. F., & Dobbert, L. (2020). Pedestrian and cyclist preferences for tree locations by sidewalks and cycle tracks and associated benefits: Worldwide implications from a study in Boston, MA. *Cities*, 106, 102111. https://doi.org/10.1016/j.cities.2018.06.024

McDonald, R., Kroeger, T., Boucher, T., Longzhu, W., & Salem, R. (2016). *Planting healthy air: A global analysis of the role of urban trees in addressing particulate matter pollution and extreme heat.* The Nature Conservancy. Retrieved April 23, 2021 from https://www.nature.org/content/dam/tnc/nature/en/documents/20160825_PHA_Report_Final.pdf

McPherson, E.G., & Muchnick, J. (2005). Effects of shade on pavement performance. *Journal of Arboriculture*, 31, 303-310.

Moreno, A., Tangenberg, J., Hilton, B.N., & Hilton, J.K. (2015). An environmental assessment of school shade tree canopy and implications for sun safety policies: The Los Angeles Unified School District. *ISPRS International Journal of Geo-Information*, 4, 607-625. https://doi.org/10.3390/ijgi4020607

Nehme, E. K., Oluyomi, A. O., Calise, T. V., & Kohl III, H. W. (2016). Environmental correlates of recreational walking in the neighborhood. *American Journal of Health Promotion*, 30(3), 139-148. https://doi.org/10.4278/ajhp.130531-quan-281

Ng, E., Chen, L., Wang, Y., & Yuan, C. (2012). A study on the cooling effects of greening in a high-density city: An experience from Hong Kong. *Building and Environment*, 47(1), 256-271. https://doi.org/10.1016/j.buildenv.2011.07.014

Nowak, D. J. (1994). Atmospheric carbon dioxide reduction by Chicago's Urban Forest. In E. G. McPherson, D. J. Nowak, R. A. Rowntree (Eds.), *Chicago's urban forest ecosystem: results of the Chicago Urban Forest Climate Project* (pp. 83-94). Radnor, PA: U.S. Department of Agriculture, Forest Service, Northeastern Forest Experiment Station.

Nowak, D. J. (2002). *The effects of urban trees on air quality.* U.S. Department of Agriculture, Forest Service.

Nowak, D. J., Crane, D. E., & Dwyer, J. F. (2002). Compensatory value of urban trees in the United States. *Journal of Arboriculture*.

Olsen, H., Kennedy, E., Vanos, J. (2019). Shade provision in public playgrounds for thermal safety and sun protection: A case study across 100 play spaces in the United States. *Landscape and Urban Planning*, 189, 200-211. https://doi.org/10.1016/j.landurbplan.2019.04.003

Parmehr, E.G., Amati, M., Taylor, E.J., & Livesley, S.J. (2016). Estimation of urban tree canopy cover using random point sampling and remote sensing methods. *Urban Forestry & Urban Greening*, 20, 160-171. https://doi.org/10.1016/j.ufug.2016.08.011

Roman, L. A., Walker, L. A., Martineau, C. M., Muffly, D. J., MacQueen, S. A., & Harris, W. (2015). Stewardship matters: case studies in establishment success of urban trees. *Urban Forestry & Urban Greening*, 14(4), 1174-1182. https://doi.org/10.1016/j.ufug.2015.11.001

Sandifer, P. A., Sutton-Grier, A. E., & Ward, B. P. (2015). Exploring connections among nature, biodiversity, ecosystem services, and human health and well-being: Opportunities to enhance health and biodiversity conservation. *Ecosystem Services, 12*, 1–15. https://doi.org/10.1016/j.ecoser.2014.12.007

Simpson, J. R., & McPherson, E. G. (2007). *San Francisco Bay Area state of the urban forest final report.* Davis, CA: U.S. Department of Agriculture, Forest Service, Pacific Southwest Research Station.

Sivarajah, S., Smith, S.M., & Thomas, S.C. (2018). Tree cover and species composition effects on academic performance of primary school students. *PLoS One*, 13, 1-11. https://doi.org/10.1371/journal.pone.0193254

Spotswood, E., Grossinger, R., Hagerty, S., Bazo, M., Benjamin, M., Beller, E., Grenier, J.L., & Askevold, R. A. (2019). *Making Nature's City.* SFEI Contribution No. 947. San Francisco Estuary Institute: Richmond, CA. Retrieved February 8, 2021 from https://www.sfei.org/documents/making-natures-city.

Spotswood, E., Grossinger, R., Hagerty, S., Beller, E.E., Grenier, J.L., & Askevold, R.A. (2017). Re-Oaking Silicon Valley: Building Vibrant Cities with Nature. SFEI Contribution No. 825. San Francisco Estuary Institute: Richmond, CA.

Stagoll, K., Lindenmayer, D. B., Knight, E., Fischer, J., & Manning, A. D. (2012). Large trees are keystone structures in urban parks. *Conservation Letters*, 5(2), 115-122. https://doi.org/10.1111/j.1755-263X.2011.00216.x

Taylor, M. S., Wheeler, B. W., White, M. P., Economou, T., & Osborne, N. J. (2015). Research note: Urban street tree density and antidepressant prescription rates—A cross-sectional study in London, UK. *Landscape and Urban Planning*, 136, 174-179. https://doi.org/10.1016/j.landurbplan.2014.12.005

Thomas, P., Bayuk, K., Samai, E., & Harrell, B. (2020). *Acterra Community-based Vulnerability Planning Pilot Project Report*. Prepared for the Office of Sustainability, County of San Mateo and Caltrans. Retrieved March 15, 2021 from https://static1.squarespace.com/static/57d1a622d1758e0dfed089fe/t/5ed039140d56e8675511258a/1590704409246/Acterra+Pilot+Project+Report+20200422.pdf

Tietje, W., Purcell, K., & Drill, S. (2005). Oak woodlands as wildlife habitat. In G. A. Giusti, D. D. McCreary, & R. B. Standiford (Eds.), *A planner's guide for oak woodlands* (pp. 15-31). University of California Agriculture and Natural Resources, Richmond, CA.

Troy, A., Grove, J. M., & O'Neil-Dunne, J. (2012). The relationship between tree canopy and crime rates across an urban – rural gradient in the greater Baltimore region. *Landscape and Urban Planning*, 106, 262-270. https://doi.org/10.1016/j.landurbplan.2012.03.010

Ulmer, J. M., Wolf, K. L., Backman, D. R., Tretheway, R. L., Blain, C. J., O'Neil-Dunne, J. P., & Frank, L. D. (2016). Multiple health benefits of urban tree canopy: The mounting evidence for a green prescription. *Health and Place*, 42, 54–62. https://doi.org/10.1016/j.healthplace.2016.08.011

Wolf, K. L. (2005). Business district streetscapes, trees, and consumer response. *Journal of Forestry*, 103(8), 396-400. https://doi.org/10.1093/jof/103.8.396

Wolf, K. L. (2007). City trees and property values. *Arborist News*, 16(4), 34-36.

Wolf, K. L., Lam, S. T., McKeen, J. K., Richardson, G. R. A., van den Bosch, M., & Bardekjian, A. C. (2020). Urban trees and human health: A scoping review. *International Journal of Environmental Research and Public Health*, 17(12), 1-30. https://doi.org/10.3390/ijerph17124371

Wood, E., Harsant, A., Dallimer, M., de Chavez, A. C., McEachan, R. R. C., & Hassall, C. (2018). Not all green space is created equal: Biodiversity predicts psychological restorative benefits from urban green space. *Frontiers in Psychology*, 9(NOV), 1–13. https://doi.org/10.3389/fpsyg.2018.02320

Ziter, C.D., Pedersen, E.J., Kucharik, C.J., & Turner, M.G. (2019). Scale-dependent interactions between tree canopy cover and impervious surfaces reduce daytime urban heat during summer. *Proceedings of the National Academy of Sciences*, 116, 7575–7580. https://doi.org/10.1073/pnas.1817561116

4 CURRENT TREE MANAGEMENT and PROTECTIONS

Overview

The City of East Palo Alto tree management program includes activities and protections pertaining to street trees, trees in other public places, and private trees. To gain an understanding of the activities involved in the City's tree management, the team met with staff from the Public Works Department and Planning Division (within the Community and Economic Department), employees from the nonprofit Canopy, and the current contracted tree pruning consultant. In this chapter, we outline the current management program, including an overview of ordinances currently in place that relate to trees, and compare the program to precedents and best practices from other cities.

Public tree management and protections

Care and maintenance of public trees

Management of the tree program is divided between the Public Works Department and the Community and Economic Development Department, Planning Division. City trees are managed with public funds and overseen by Public Works whereas the Planning Division manages the bulk of the private tree protections through the evaluation and review of tree removal permits.

The care and maintenance of public trees is the responsibility of the Public Works Department, which primarily performs an administrative role. The total budget of the public tree program is $150,000, with approximately $30,000 allocated for staff time and $120,000 budgeted for contract services. Personnel funded for tree care includes two maintenance staff and the Maintenance Division Manager. Due to limited staff availability, pruning and removals are performed on an as-needed basis mainly under contracted services. This is considered a "reactive" rather than a "proactive" approach to tree management, the latter of which would include a program of inspection, scheduled pruning cycles, removals, and replacement. Consequently, the response to outside requests for pruning and removal is administrative and only reactive inspections and minor tree work are accomplished with City resources.

Service requests for tree maintenance are received from the public via the City's website through their internal iWorQ work management system. In responding to requests, the existing tree inventory is used to determine if the tree is a public or private tree (for more details on the tree inventory, see Chapter 3). The tree inventory is available to the City through an ArborPro database, which can be accessed by City staff. As the City does not employ an in-house arborist, after confirming tree status, the Maintenance Division Manager conducts a field inspection to confirm the reported disposition of the tree(s) and determine the appropriate course of action to be taken. Most tree pruning needs and all removals are turned over to contract maintenance service for resolution. None of this work is recorded in the ArborPro tree inventory. Public Works employees who regularly access the tree inventory have noted that there are several areas of public trees missing from the current data.

Tree care statistics

Information from service requests, budget reports, and monthly logs was compiled to provide a snapshot of maintenance activities. The following data reflects annual average tree-related maintenance activities completed and program expenditures over the past three to five years.

- 100-120 trees planted by Canopy volunteer efforts
- 100 trees pruned by an on-call tree service
- 5-10 trees removed by an on-call tree service
- 40 tree removal permit applications processed
- $120,000 budgeted for tree contract services
- Expenditure per tree serviced was:
  - $95 for a tree with a 0-6" diameter*
  - $150 for a tree with 7-12" diameter*
  - $250 for a tree with 13-18" diameter*
  - $300 for a tree with 19-24" diameter*
  - $550 for trees > 24" in diameter*

*Measured at 54" above grade

Tree planting and community participation

A sustainable urban forest requires diversity in tree species composition and age, and a planting program that can keep pace with tree removals. Tree planting is not a planned element in the City's tree management program. For the last decade, the Palo Alto-based nonprofit Canopy has provided all of the tree planting services for the City.

In 2005, Canopy launched a multi-year initiative to plant at least 1,000 trees in East Palo Alto. In collaboration with City leadership, and to address air pollution from US 101 Highway, Canopy identified priority planting zones along the four stretches of the highway on East and West Bayshore Road, totaling 2.1 miles of roadway between Embarcadero Road and Willow Road. These efforts have contributed to the planting of over 3,000 trees in East Palo Alto. Canopy's planting efforts in the city continue, largely funded through Cal Fire grants and Canopy's fundraising efforts, and carried out with Canopy staff and volunteers.

Nobel Peace Laureate Wangari Maathai with mayor Ruben Abrica, former mayor Pat Foster, and members of the community. Photo: Canopy.

Canopy also leads current work to involve the community in urban forestry. In 2007, Canopy created the Teen Urban Foresters (TUF) program to involve high school students from various high schools in East Palo Alto. The program offers part-time employment to underserved youth to encourage leadership and tree stewardship. Youth staff assists with growing trees from seedlings, tree planting, tree maintenance (including irrigation systems), and creation of public outreach campaigns, among other programs. The TUF program continues to employ 8-16 youth every year.

This type of partnership is not uncommon. In many cities, local community organizations or nonprofits take on much or all of the tree planting responsibilities. This model is used locally in San Jose with Our City Forests and in San Francisco with Friends of the Urban Forest, and Canopy provides these services for the Cities of Menlo Park, Mountain View, and Palo Alto as well as East Palo Alto and unincorporated areas of Redwood City. Cities may also partner with multiple organizations for work on the urban forest. For example, the City of Los Angeles partners with several different organizations including Koreatown Youth and Community Center, Los Angeles Conservation Corps, TreePeople, the Hollywood Beautification Team, and North East Trees.

Codified public tree protections and maintenance

The City does not have a comprehensive stand-alone tree ordinance within the municipal code. Rather, there are two ordinances that contain provisions for activities regarding public trees; one addressing tree pruning in Title 12: Streets, Sidewalks and Public Places, and the other in Title 13: Public Services, which provides specifications for tree planting.

Title 12 encompasses requirements for park and public space use under the coordinated umbrella of the Public Works Department and the Planning Division and includes Chapter 12.16 "Cutting and Trimming Trees on City Streets and Public Places." This chapter focuses on regulations including the conditions, prohibitions, and permissions required to trim or remove any tree located in the public right-of-way or in a public place, or any private tree overhanging a public right-of-way or public place. Written consent for any person other than City staff to perform these activities is required from the Public Works Director for public trees. The Public Works Director has the authority to override the refusal of a private property owner.

Title 13, Chapter 13.24, Article VII. Section 13.24.410 "Street Trees" gives authority to the Public Works Director to perform tree planting functions for development projects with standard detail specifications for the planting of street trees including:

- Container size, tree spacing, and distances from utilities such as light poles, fire hydrants and sewer lines.

- Planting standards for basin size, staking, backfill, and tree ties.

- Standard landscaping areas in parking lots and around model homes.

Planting standard details for the purpose of safety and maintenance specify that trees should be at least:

- Twenty feet from curb returns at street intersections.

- Ten feet from light standard poles.

- Ten feet from fire hydrants.

- Five feet from sidewalks, driveways, buildings, walls and permanent structures.

- Ten feet from water and sewer lines.

The public tree ordinances provide staff with direction to follow in response to tree service requests. Currently, if it is determined that service is required to remove a dead or declining public tree, one of two approaches are typically pursued:

- The Maintenance Division Manager forwards the request to an on-call tree service company who completes the work and maintains a record of their service call.

- As per Chapter 12.16.020, the requestor must obtain written consent from the Public Works Director for approval to prune or remove the tree in compliance with the provisions of the chapter. If the request pertains to a private tree that overhangs public property, written consent must first be obtained from the private property owner. If the property owner denies consent, the requestor can apply for written consent from the Public Works Director.

East Palo Alto Municipal Code definitions

Private trees: Trees located on commercial, industrial, or residential property outside the boundaries of public rights-of-way, easements, and public places.

Protected trees: Defined in Section 18.28.040 as (1) any private tree with a trunk circumference of 24 inches or greater at a height of 40 inches above grade, (2) any public tree within a public street or right-of-way regardless of size, (3) trees planted and/or preserved as a condition of development, and (4) any tree planted as a requirement for an unlawfully removed tree.

Public trees: Street trees and trees located on other public properties such as in parks, along trails, in landscaping around City offices and within public parking lots.

Remove: Tree removal includes (1) complete removal, such as cutting to the ground or extraction, or (2) taking any action foreseeably leading to the death of a tree or permanent damage to its health, including but not limited to: excessive pruning, cutting, girdling, poisoning, over-watering, under-watering, unauthorized relocation or transportation, or trenching, excavating, altering the grade, or paving within the dripline. (per Chapter 18.28.040)

Street trees: Trees located within the right-of-way or easement along every street within the city.

Tree:

- Any fruit, shade, ornamental or other tree of any kind or nature (per Chapter 12.16)

- A woody perennial plant characterized by having a main stem or trunk, or a multiple-stemmed trunk system, with a more or less definitely formed crown, and usually over 10 feet high at maturity. This definition does not include trees planted, grown, and held for sale by licensed nurseries or the first removal or transplanting of trees as part of the operation of a licensed nursery business. (per Chapter 18.28.040)

Trim: the cutting off or removal of any limbs or branches of trees (per Chapter 12.16)

How do public tree management and protections compare with other cities?

In order to understand how tree management practices and protections in East Palo Alto compare to other municipalities, we conducted a benchmarking survey of urban forest programs in ten local cities (Table 4.1). Selected cities shared one or more commonalities with East Palo Alto, including factors such as climatic conditions and population size, or were adjacent neighboring cities with common borders. For each City in the benchmarking survey, questions about tree care operations and ordinance regulations were answered by phone survey, from urban forest master plans, or from department operational data. The municipal codes of each City were reviewed for ordinance provisions regarding both public and private trees. Findings from our local survey were compared with results of a published 2014 survey of municipal tree management practices from 667 cities across the US (Hauer and Peterson, 2016).

Table 4.1. Characteristics of cities used to benchmark East Palo Alto's tree program and determine best practices.

	Population	Number of Public Trees	City Tree Ordinance?	Private Trees Protected?
Atherton	7,187	1,565	Yes	Yes
Atwater	28,168	7,922	Yes	Yes
East Palo Alto	29,314	5,745	Yes	Yes
San Pablo	30,000	Unknown	Yes	Yes
Foster City	32,000	Unknown	No	No
Menlo Park	34,698	19,500	Yes	Yes
Palo Alto	65,364	46,000	Yes	Yes
Richmond	110,567	22,000	Yes	Yes
Berkeley	121,485	38,000	Yes	Yes
Vallejo	121,692	53,000	Yes	Yes
Hayward	159,203	30,000	Yes	Yes

East Palo Alto currently has 13.5% tree canopy cover which is below the national average of 27%. Other benchmark cities with low tree canopy cover include Richmond (12.7%), Atwater (9%) and Hayward (1% of total land area; 23% of sidewalk area covered). Richmond has a stated goal to reach 15% by 2020, and 25% by 2030. Atwater has identified a tree canopy cover potential for the city of 31%. Nationally, the mean tree canopy cover goal is 44% (Hauer & Peterson, 2016).

Ordinances regulating public trees are very common among cities. Nine of ten local benchmark cities surveyed have a street tree ordinance that regulates public trees to varying degrees, with seven of the cities having a robust street tree ordinance. This reflects national trends, where over 90% of cities have at least one tree ordinance (Hauer & Peterson, 2016). Menlo Park, Vallejo, Richmond, and Atwater specifically refer to their tree management or tree master plan as governing tree maintenance operations. The ordinances consistently identify trees in the street, right-of-way, and public places as City trees, with policies that range from maintenance responsibility to protection, permits, violations, and appeals.

Nationally in cities that are similar in size to East Palo Alto, Public Works departments are most often responsible for public trees, with Parks and Recreation, Forestry, Urban Forestry, or Planning departments sometimes taking responsibility (Hauer & Peterson, 2016). 79% of these similar cities have the staff responsible for tree management attend tree care workshops, while only 13% provide no formal training. 60% have an International Society of Arboriculture (ISA) certified arborist on staff. On average, these cities employ 4-5 full time-equivalent staff members as part of public tree management programs.

Seven of the ten benchmark cities surveyed maintain trees In the public right-of-way. Similarly, 64% of cities nationally are responsible for trees in the right-of-way (Hauer & Peterson, 2016). Exceptions include Atherton, where most trees in the public right-of-way are private trees and the responsibility of the property owner. The Town maintains a list of tree care providers as a resource for property owners to promote the use of best management practices. They also have a heritage tree ordinance that applies to trees over 15.2" in diameter with standards and specifications for their care and protection. In Vallejo, residents are responsible for maintenance activities for City trees on their property. A permit is required for trimming, pruning and removal approval. In the case of tree removal, a permit fee is required at the time of application and used by the City to purchase and plant a replacement tree on the property.

Five benchmarked cities prune City trees on a regularly set pruning cycle. However, Berkeley and Foster City, like East Palo Alto, prune on an as-needed basis to clear obstructed views of blocked streetlights and signs, usually in response to a request for service. The as-needed approach is reactionary, managing problems as they arise. Responding to service requests can be disruptive to maintenance schedules and more costly if contracted service-on-demand is required. Conversely, proactive maintenance can prevent issues from developing by projecting maintenance needs based on past activity and setting priorities.

In eight of ten benchmarking cities, tree pruning and removals either were contracted out entirely or were performed by a combination of in-house and contracted crews. With a hybrid arrangement, assistance was most often contracted for removals and emergency response.

Budgets allocated for tree management vary. Nationally, the mean annual tree management budget for cities similar to East Palo Alto in size was about $340,000 in 2014 (Hauer & Peterson 2016). Given its location in one of the most expensive regions of the US, East Palo Alto might be expected to require a higher budget. About 86% of cities nationally rely on the General Fund to support tree maintenance activities. Among local benchmarking cities, annual budgets for tree management ranged from $1.1M in Hayward to $2.4M in Palo Alto. For smaller-sized cities such as Foster City, which only maintains trees in medians and parkways, and Atherton, which does not prune trees in the right-of-way, budgets ranged from $90,000-95,000 respectively, slightly lower than but comparable to East Palo Alto at $150,000.

Table 4.2: Tree management program information for ten local benchmarking cities.

City	Has tree inventory?	Has urban forest master plan?	City maintains right-of-way trees?	Pruning approach	$ budgeted per City tree	% tree canopy cover
Atherton	Yes	No	No	N/A	Unknown	Unknown
Atwater	Yes	Yes	Yes	Proactive	Unknown	9%
East Palo Alto	Yes	In progress	Yes	As needed	$26	13.5%
San Pablo	No	Yes*	Yes	Proactive	Unknown	Unknown
Foster City	No	No	Yes**	As needed	Unknown	Unknown
Menlo Park	Yes	No	Yes	Proactive	$74	26%
Palo Alto	Yes	Yes	Yes	Proactive	$71	38%
Richmond	Yes	Yes	Yes	Proactive	Unknown	12.7%
Berkeley	Yes	No	No	As needed	$65	22%
Vallejo	Yes	In progress	No	Proactive	Unknown	13%
Hayward	Yes	Yes***	Yes	Proactive	$51	1%

*Master Landscape Plan contains tree provisions for planting, maintenance and protection
**In medians only
***Resource Analysis includes the existing nature of the urban forest.

Private tree protections

Codified protections for private trees

The City of East Palo Alto has protected trees on private property since July 17, 1989, showing a commitment to tree protection. This commitment is a good foundation for supporting the City's urban forestry goals. Currently, the City protects trees on private property through Chapter 18.28 - Landscaping and Trees, within the Development Code. The goal of the code is to carry out the policies of the City's Vista 2035 General Plan (City of East Palo Alto, 2017). Authority for carrying out the provisions of the code resides with the Community and Economic Development Director. Chapter 18.28 states support for policies regarding the use of landscaping and trees, including preserving trees for the health, safety, and welfare of the community, and using trees to preserve scenic beauty, prevent erosion of topsoil, protect against flood hazards, counteract pollutants in the air, maintain climatic balance, and decrease wind velocities.

Preservation measures regulating the removal of protected trees are in Section 18.28.40. Protected trees are defined as

1. Any private tree with a trunk circumference of 24 inches or greater at a height of 40 inches above grade;

2. Any public tree within a public street or right-of-way regardless of size;

3. Trees planted and/or preserved as a condition of development; and

4. Any tree planted as a requirement for an unlawfully removed tree.

The ordinance also requires tree protection during construction.

A tree removal permit is required if a tree owner wants to remove a protected tree, unless an exemption applies. To secure a permit, the property owner must obtain an arborist report to show that the tree or trees meet one or more criteria for removal laid out by the ordinance:

1. The tree is dead or infected with a terminal disease.

2. The tree is structurally unsound and cannot be corrected or the risk cannot be significantly reduced by traditional pruning, cabling, or bracing.

3. The tree is causing visible damage to property, which cannot be corrected without destroying the tree canopy or root system.

The applicant submits the arborist report along with a tree removal permit application and tree removal permit fee as specified in the City's fee schedule. The most recent Comprehensive Fee Schedule adopted on July 1, 2021 sets the permit fee for removal of one or two trees at $356 per tree and removal of three or more at $762 per tree (City of East Palo Alto, 2021). The Planning Division then reviews and processes tree removal applications to determine whether the tree meets the criteria for removal.

Property owners are required to notify abutting property owners and tenants of their intent to remove a protected tree both when the permit application is submitted and again 48 hours prior to removal. Removals can be appealed, and permits can be denied if the Planning Division determines that the tree provides particularly important benefits to the community, contributes substantially to the appearance of an area, or is part of a mutually dependent group of trees. The ordinance does not state who can make an appeal. If the permit is approved, the tree owner is required to either plant one or more replacement trees or pay an in-lieu fee, as determined by staff reviewing the application.

There are four situations delineated by the ordinance in which a tree removal permit is not required for removal of a protected tree.

1. Emergencies, where a tree presents an immediate hazard to life or property.

2. Trees on public property may be removed by City employees where necessary for safety.

3. Public utilities may remove trees to comply with safety regulations.

4. Projects where the removal of trees has been authorized as part of a development approval granted by the City.

The fourth situation exempts development projects with a design review permit from filling out an application for tree removal, as the review is included in the total project review. These applicants do not follow a normal tree removal process.

If a protected tree is removed without a permit in any other circumstances, penalties can include replacement on site with three or more new trees or a mitigation fee where on-site replacement is not feasible, as determined by staff. The City can also revoke the business license of any person or business who violates the tree protection ordinance with an unpermitted removal. A tree violation, including unpermitted removals or trimming, is considered an administrative penalty, and the cost is set at no less than $100 by East Palo Alto Municipal Code 1.14.030. If a constituent receives a violation for the removal of a protected tree, they are required to retroactively apply for a tree removal permit. Through this process, they are required to pay both the penalty and the permit application fee. Violations are processed by code enforcement officers, though community members perceive little enforcement of current tree protections (see Chapter 5).

Each year, the City processes about 40 tree removal permit applications. About half of these applications are typically for trees that have already been removed. An unknown number of protected trees are removed by development projects approved through the design review process.

How does East Palo Alto's private tree protection compare to other cities?

Private tree ordinances in other local benchmarking Cities were either stand-alone ordinances that designate heritage trees by size and species (Palo Alto, Menlo Park, Atherton, Berkeley, San Pablo, Hayward), or were provisions within street tree ordinances (Atwater) or other development ordinances to protect private trees during construction (Richmond). Seven of ten benchmarking cities compared had specific criteria for tree protection in their ordinances, including protecting any tree with a trunk diameter of greater than 6" (San Pablo), specific native trees with a trunk diameter greater than 4" and all trees with a trunk diameter greater than 8" (Hayward), or all trees that are larger than 25' height and diameter of 10" (Vallejo).

Eight of ten benchmarking Cities surveyed required a permit to remove a protected tree. Permit fees ranged from $54 in Vallejo on the low end to $662 in San Pablo on the high end. With a fee of $356 for 1-2 trees, East Palo Alto is comparable to the average fee of $312 and is above the high end at $762 each for 3 or more trees. Outliers included Berkeley which does not charge a fee, and Atherton with a fee of $2,100 for the removal of a designated heritage tree. Fees were generally included in the City's master fee schedule and approved by city council, as in East Palo Alto.

Arborist reports are a requirement for a tree removal permit in most Cities. Six of ten benchmarking Cities surveyed required an arborist report to accompany the tree removal permit request. In Atwater, City staff prepare a tree report for review by the Community Development and Resources Commission rather than requiring an independent arborist assessment.

Criteria for private tree removal were similar across Cities. Eight of ten benchmarking Cities surveyed cited criteria that would qualify a tree for removal if it is dead or hazardous, has high probability of failure, impedes progress or vision, overhangs the street, causes infrastructure damage such as sidewalk raising (Hayward), or hinders reasonable use or enjoyment of the property (Atherton). These criteria are similar to the current removal criteria in East Palo Alto.

Which trees are protected?

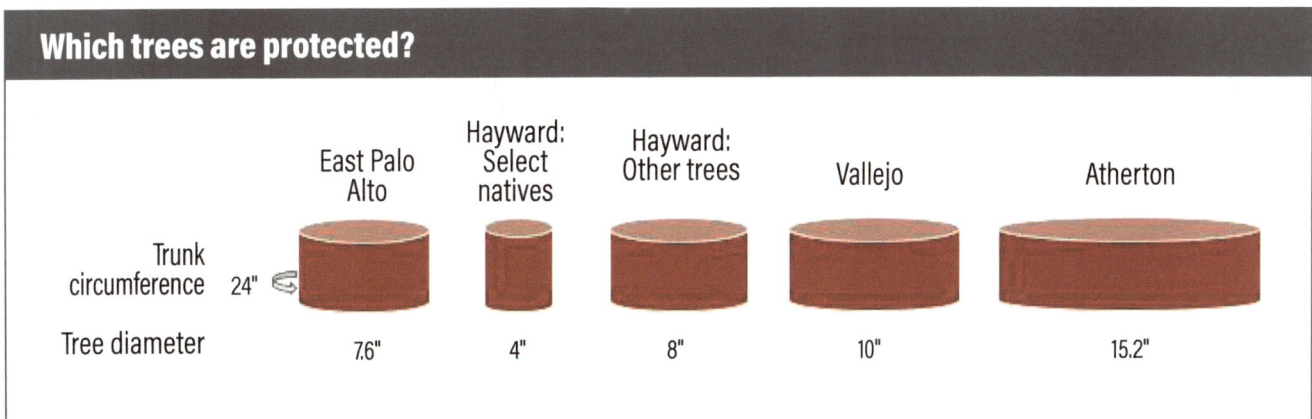

	East Palo Alto	Hayward: Select natives	Hayward: Other trees	Vallejo	Atherton
Trunk circumference 24"					
Tree diameter	7.6"	4"	8"	10"	15.2"

Seven of ten benchmarking Cities had an appeals process associated with tree removals. Appeals were submitted to and reviewed by the department issuing the permit, generally the Planning Department. In two Cities, appeals were directed to a community advisory board such as the Beautification Advisory Board (Vallejo) or the Planning Commission (Atherton).

Conclusions

Unlike neighboring cities, East Palo Alto's public tree management program is not well defined by a robust public tree ordinance. The City lacks a comprehensive street tree ordinance that designates responsibility and authority to make decisions and establish policies for tree planting, maintenance, preservation, and removal. The current maintenance program responds to emergency needs, but is largely administrative and unable to proactively address tree care needs. Canopy is a key partner in East Palo Alto, raising awareness of the value of trees and supporting the City with tree planting.

Ordinance provisions for protection of trees on private property are largely in alignment with benchmark Cities. However, the code does not use industry standards for some criteria, and application review and enforcement are difficult. Protecting trees within the scope of available resources will be a continuing challenge, which can be made easier by clarifying the criteria, scope, and process for tree removal permitting.

Volunteers holding fresh mulch. Photo: Canopy.

References

City of East Palo Alto. Title 12, Chapter 12.16 - Cutting And Trimming Trees On City Streets And Public Places. Retrieved January 25, 2022 from https://library.municode. com/ca/east_palo_alto/codes/code_of_ordinances?nodeId=TIT12STSIPUPL_ CH12.16CUTRTRSTPUPL

City of East Palo Alto. Title 13, Chapter 13.24, Article VII. Section 13.24.410 - Street trees. Retrieved January 25, 2022 from https://library.municode.com/ca/ east_palo_alto/codes/code_of_ordinances?nodeId=TIT13PUSE_CH13.24WASY_ ARTVIILAIRST_13.24.410STTR

City of East Palo Alto. Title 18, Chapter 18.28 - Landscaping and Trees. Retrieved January 25, 2022 from https://library.municode.com/ca/east_palo_alto/codes/code_of_or dinances?nodeId=EAPAALDECO2018EDCUORNO09-2020ADDE152020_ TIT18DECO_ART3REAPALZO_CH18.28LATR

City of East Palo Alto. (2010). Title 1, Chapter 1.14, Section 1.14.030 - Amount of administrative penalties. Retrieved January 25, 2022 from https://library. municode.com/ca/east_palo_alto/codes/code_of_ordinances?nodeId=TIT1GEPR_ CH1.14ADRE_1.14.030AMADPE

City of East Palo Alto. (2017). *Vista 2035 East Palo Alto General Plan*. Retrieved May 10, 2021 from https://www.ci.east-palo-alto.ca.us/econdev/page/general-plan-2035- east-palo-alto

City of East Palo Alto. (2021). *City of East Palo Alto Comprehensive Fee Schedule*. Retrieved January 25, 2022 from https://www.cityofepa.org/sites/default/files/ fileattachments/finance/page/4461/east_palo_alto_comprehensive_fee_schedule_ eff.01.2021_U.pdf

Hauer, R., & Peterson, W. D. (2016). *Municipal Tree Care and Management in the United States: A 2014 Urban & Community Forestry Census of Tree Activities*. Special Publication 16-1, College of Natural Resources, University of Wisconsin – Stevens Point.

5 COMMUNITY PERSPECTIVES

Approach

As part of the Urban Forest Master Plan development process, the authors engaged with over 350 community members to understand community perspectives on trees and to gather input about urban forestry needs in the city. The engagement approach included several different components.

- **Steering Committee:** A standing group of community members participated throughout the process in goal setting, vision setting, community engagement interpretation, and policy recommendations.

- **Online Survey:** A public survey was online during the summer of 2021 to gather input from the community on attitudes toward trees, perceived benefits and problems, and tree protection and planting efforts.

- **Project Website:** Information about the project was made available in a dedicated project website, which was linked from the City's planning website. The project website also hosted the online survey.

- **Focus Groups:** A mix of online and in-person, discussion-based meetings were held to gain perspective from specific groups of stakeholders.

- **City Meetings**: The project team attended public City Council, Planning Commission, and Senior Advisory Committee meetings to share updates on project progress and planned outcomes.

- **Tablings:** Informational booths at community events aimed to spread awareness and gain participation in the online survey.

- **Interviews:** Conversations with individual key stakeholders helped gather detailed insights into potential opportunities, barriers, and needs for the urban forest.

Steering committee

The project steering committee was composed of nine members, including residents, representatives from City bodies, and other key stakeholders. This group met four times over the course of the project to review and comment on project materials.

Online survey and project website

To gain broad feedback from the community on attitudes and priorities around trees in East Palo Alto, we distributed an online survey. The survey was available from April 28 through August 18, 2021 and was provided in English, Spanish, Samoan, and Tongan. The survey was available through the project website, and was shared via the City newsletter, posts and newsletters from community organizations including Nuestra Casa, Canopy, Fresh Approach, and YUCA, and social media posts on Facebook and Nextdoor. All survey respondents were entered into a raffle for Visa gift cards and a fruit tree.

We received a total of 208 unique responses to the online survey. 92% of respondents completed the survey in English, and 93% lived in East Palo Alto.

To understand the ability of these survey results to represent the community of East Palo Alto, we compared reported survey demographics with US Census data. Survey respondents broadly matched the overall city composition by age, income, and gender. 45% of survey respondents were Hispanic/Latino, compared to 66% from census demographics. The survey somewhat overrepresented homeowners compared to renters (53% of respondents were homeowners, compared to 40% from census data), and respondents were more highly educated than city demographics, with 83% of respondents having at least some college experience compared to 42% of residents overall.

The survey included three free response questions:

1. In a few words, tell us what you like or dislike about trees in your neighborhood.

2. Are there other benefits or problems associated with trees that you think are important? If so, please tell us here.

3. If you have other ideas, thoughts or questions about how the City is or should be investing in or protecting trees, please share them here.

Responses to these questions are used in the following sections to give exemplary quotes relating to different themes.

Canopy's Teen Urban Foresters planting a lemon tree given as a prize to a survey respondent. Photo: Canopy.

Focus groups, city meetings, and tabling events

To gain a more detailed understanding of the priorities and concerns around tree planting, we held five focus group meetings with community members. In these meetings, we asked participants to explain their attitudes toward tree planting and any barriers they may face. Some groups also responded to specific prompts. Some meetings were held virtually and some in person due to shifting public health considerations and group preferences. Participants who were not public employees were compensated for their participation.

Focus groups were organized with the following themes:

1. Youth, hybrid meeting - included discussions of how East Palo Alto could look in the future and development of the vision statement (5 student participants)

2. Polynesian and Latinx communities, in-person meeting (15 participants)

3. Faith community, in-person meeting (4 participants)

4. Business and development community, hybrid meeting (9 participants)

5. Residents who have applied for tree removal permits, virtual meeting - focused on the permitting process (3 participants)

We also attended a City Planning Commission meeting on May 24, 2021 and a Senior Advisory Committee meeting on October 6, 2021 to share project information and gain perspective from these groups. Additionally, we presented at City Council meetings on June 15, 2021 and November 16, 2021 to share information and gain feedback from the Council and the public.

While public health considerations limited in-person engagements, the team was able to interact with nearly 100 residents in person through several events. These included tabling at a local farmers market, a community open house, an art show, and a shopping center. Tabling engagements included talking with passers-by to share information and encouraging participation in the online survey. In addition, we held a pop-up event on July 31, 2021 to provide information and hold conversations with interested residents.

Interviews

In addition to the above community-focused engagement efforts, we also held individual conversations with 18 key stakeholders to better understand their perspectives and priorities around tree planting. These conversations included City staff, land owners, and community leaders.

Community themes

We identified seven themes emerging across our community engagement efforts.

> "The **more trees** the better! There just aren't enough."

> "Some of my neighbors have **big trees that I love**, but most houses don't have trees at all, or not many!"

> "I **love the green** our trees add to the landscape. I dislike that there are **so few**."

1. Strong commitment to stewarding the land and growing the urban forest

Overall, the community members we spoke with were enthusiastic about growing East Palo Alto's urban forest.

Community members responded strongly to tree canopy cover maps showing the disparity between tree canopy cover in East Palo Alto and neighboring cities, remarking that these maps looked similar to other maps of environmental hazards and socioeconomic disadvantage in the area. Upon seeing the low existing tree canopy cover, nearly all community members we spoke with expressed a wish to see more trees planted to reduce the gap between East Palo Alto and its neighbors.

Many residents felt that trees in the city had declined over the time they had lived there, and recalled experiences of large trees shaping community character in the past. Community members felt ties to historic agricultural land uses in the area as well as personal connections to caring for the land, and saw urban forest protection and growth as a way to provide stewardship to the local environment. Trees were also seen as a way to connect with cultural roots, both indigenous local cultures and the cultures of current immigrant residents.

40 of the 208 survey respondents wrote that the city or their neighborhood needed more trees.

Community members were interested in finding ways to support tree planting efforts through existing community groups and grassroots action.

Community youths tending to trees. Photo: Canopy

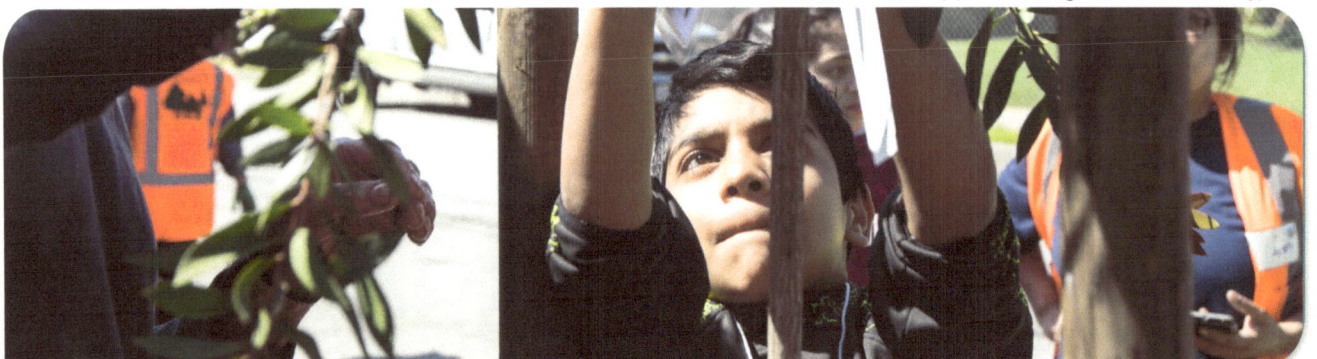

2. Need for better communication and partnership between the City and the community

Residents expressed a desire to pursue community-based strategies to expand the urban forest, rather than relying on action from the City. Both City staff and residents emphasized the need for action to originate from and be driven by the local community, rather than external partners.

One barrier that was identified through focus group conversations was the lack of materials and guidance around tree planting, maintenance, and removal in the city for non-English speakers. Residents noted that City tree removal permitting guidance and materials were not available in commonly spoken languages such as Spanish and Pacific Islander languages, and that outreach efforts often were conducted only in English.

With much of the city made up of privately owned land, private landowners and residents are a key part of the path toward growing the urban forest, and improved communication and collaboration is essential.

3. Positive attitudes toward trees and the services they provide

Overall, people in the community like trees and see value in having them in the city. Only 2% of survey respondents said they disliked trees, and comments about trees were largely positive.

When asked about the potential benefits and drawbacks of having trees, respondents overall ranked benefits as more important than problems, showing the high perceived value of trees (Fig. 5.1). The most highly rated benefits of trees from survey respondents were air quality improvements (94% rated very important) and heat mitigation and shading (89% rated very important). These benefits were also mentioned by focus group participants, and are tied to health outcomes for residents. Many survey respondents mentioned birds and providing places for wildlife as a benefit of urban trees. This was supported by 57% of respondents who said that whether or not a tree was native to the region was a very important trait to consider when choosing a new public tree. Residents also often brought up enjoyment of fruit trees and the feelings of community from gathering and sharing fruit during focus group conversations, and 41% of survey respondents considered fruit-bearing to be a very important trait to consider for public trees.

"I love that **they provide habitat** for all kinds of life. They keep my apartment nice and **cool in the summer**. The [sight] and sound of them blowing in the wind is **comforting**."

"They are beautiful. **Provide habitat** for birds and wildlife. They **dampen the sound** of the highway and **provide cleaner air** given how close we live to the freeway."

Increasing property values was the lowest-ranked benefit of trees for this community, with 46% of respondents ranking it as very important, and 23% considering it to be not at all important.

Negative aspects of trees were also mentioned, but overall were considered to be less important than the benefits. The biggest concern was for their potential to cause root damage to underground utilities, foundations, or walkways, which 39% said was very important and an additional 50% said was slightly important. Conflict with overhead utility wires and the cost of maintenance, pruning, and removal were the next greatest concerns, with 83% of respondents very or slightly concerned about utility conflicts and 73% very or slightly concerned about costs.

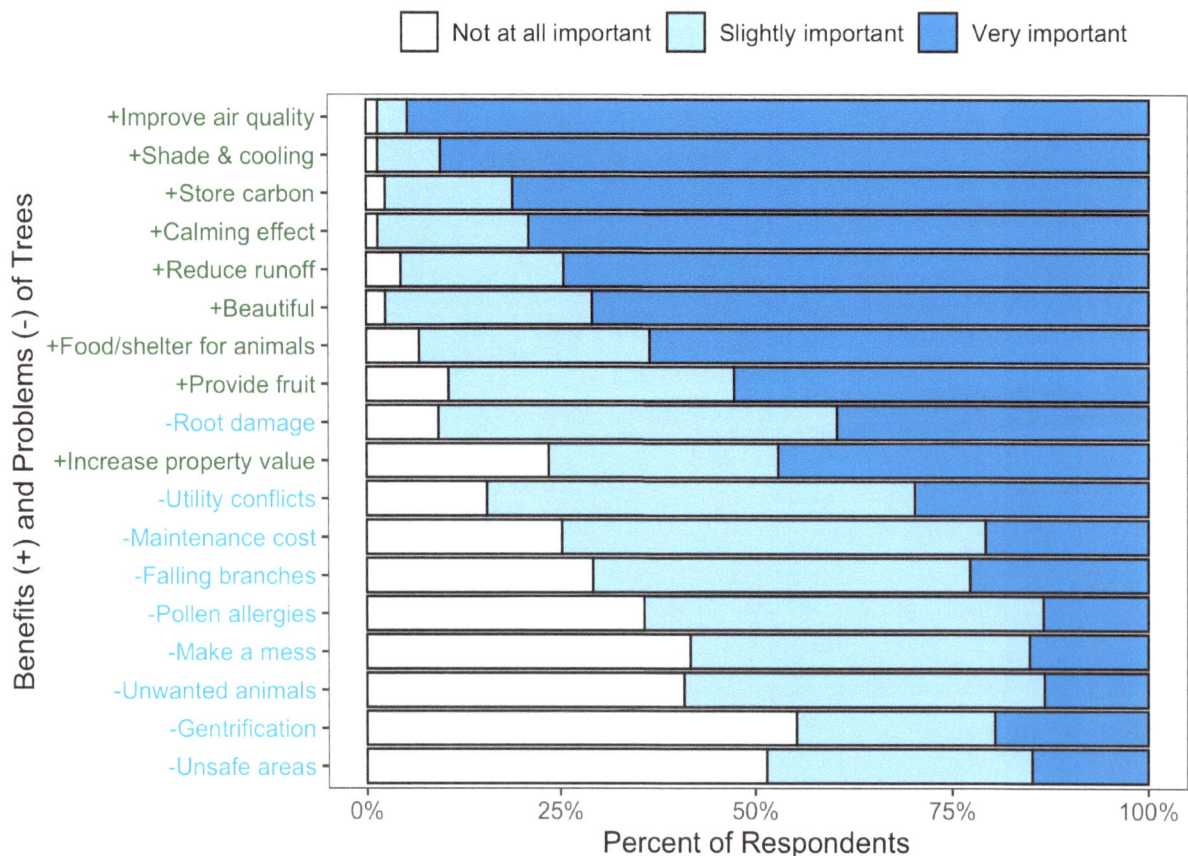

Figure 5.1. Responses from 208 community members to two survey questions. The first asked how important each of several benefits of trees are for East Palo Alto, considering trees planted in parks, along streets, and in private yards. The second asked which of the listed problems associated with trees respondents were concerned about.

4. Concern over tree maintenance and management

The top concern mentioned by survey respondents was tree maintenance, with 20% of respondents bringing up maintenance concerns in at least one of their three free response answers. Community members were concerned about improper pruning and maintenance, particularly of City trees, which could lead to increased safety risk and poor tree health. They also saw maintenance needs as an issue for residents, particularly lower-income households and seniors on fixed incomes who may be less physically or financially able to provide necessary tree maintenance to prevent hazards and maintain tree health.

Conversations with landowners and City staff also cited maintenance as a key concern, and in particular raised questions about how to fund continued maintenance of trees as they grow older and larger. While some landowners saw the benefits of trees and wished to plant more, planning for future maintenance was seen as a major barrier to growing the urban tree canopy. Landowners, homeowners, and City staff felt that they could access new trees for planting through Canopy and appreciated the young tree care provided over the first 3-5 years after planting, but maintenance following this initial period was a serious concern. In particular, East Palo Alto schools working with limited landscaping budgets found continuing tree maintenance to be the primary barrier to more tree planting on school grounds.

"**Not enough trees** and the ones we do have are not well taken care of."

"Maintenance plan, **make sure they are being trimmed on a schedule** and not just when complaints are received"

"The **City does not maintain our trees or meadows properly**. Many trees that have been taken out because of drunk drivers or speeding are not replaced. Why?"

"El peor problema es que el vecindario y **la ciudad no mantienen y cuidan nuestra ciudad**. Y cada vez que arreglan al poco tiempo dejan secar y destruir, y se tira todo lo que se invierte y se trabaja." (Translation: The worst problem is that the neighborhood and the City do not maintain and take care of our city. And every time they fix it in a short time they let it dry and destroy, and everything that is invested and worked for is thrown away.)

5. Concern for future resilience, adaptation to climate change, and water availability

81% of survey respondents considered newly planted public trees' ability to adapt to future climate change to be very important, and 16% said it was slightly important. In particular, water availability is a major concern for residents, who felt constrained by both current and possible future water restrictions. People were concerned that future drought would lead to restrictions that would prevent tree irrigation, and felt that adding more trees might lead to unsustainable water costs.

6. Concern about knowledge, enforcement, and process of the existing Tree Protection Ordinance

Many residents were not aware of the existence of a tree protection ordinance in East Palo Alto, with 10% of survey respondents saying they thought the City did not have such an ordinance, and 50% saying they didn't know.

Residents who had experienced applying for a tree removal permit described different pathways with varying information, cost, and difficulty for obtaining a permit. Applicants worked with different staff members and were provided with different materials to support their applications. The process was sometimes long, with unexpected requirements and costs. Survey write-in responses echoed these difficulties, with complaints that the current process is too complicated and expensive.

Survey respondents and focus group participants frequently shared stories of unpermitted tree removals and lack of enforcement, and expressed frustration that the existing tree protection ordinance was not very effective.

Community members expressed concern that for many households, the cost of applying for a tree removal permit and hiring an arborist to submit the associated report would be prohibitively high. Survey respondents and focus group participants agreed that the costs for permitted tree removals likely discouraged people from following the legal process.

Some expressed a desire for greater flexibility for private property owners, with developers and City trees more firmly controlled by tree protections. 57% of survey respondents felt that the City should protect trees on private residential property, and 75% thought the City should protect trees on commercial property.

"There are trees that are REMOVED on the weekends, undercover [or] absent of discovery? Give out **rewards for the reporting of trees being chopped down**. Also the rule of the tree being 41 inch in diameter doesn't protect the trees that mature that can't reach that size. For instance "palm trees". **Corporate developers don't care about trees**."

"People are always doing **illegal tree trimming or cutting** them down on the weekends when nobody is checking. Don't police people unnecessarily, but have someone available or a **hotline to call**"

"I'm **concerned about lower income property owners** being able to afford to legally deal with trees."

"While I believe trees on private property should be protected as well, I do worry about the **cost burden for residents** that may need to remove trees. In order to enforce the ordinance on private properties, there should be a source of funds that can **provide financial support**"

7. Desire for developers to be responsible for additional planting

With redevelopment occurring in the Ravenswood Employment District in particular, some community members hoped redevelopment efforts would create opportunities for more tree planting and investment in the local environment. Members of the business community supported these wishes, but were concerned about the feasibility of tree planting in the Ravenswood Employment District due to possible saltwater intrusion and poor physical conditions. Community members were interested in the possibility of supporting in lieu planting, either physically planting or providing funds for planting and maintenance of trees outside the development area.

Survey respondents largely agreed that new construction and redevelopment projects should be required to protect existing trees (79%) and to plant new trees (75%).

Planting trees. Photo: Canopy.

Conclusions

East Palo Alto residents at large are committed to growing and stewarding the urban forest, and they believe in this work being driven through grassroots initiatives and community empowerment. While the community generally feels that the benefits from trees outweigh the costs, these costs, particularly related to tree maintenance, have real impacts on the health of trees and individuals' willingness to plant a tree.

While the community wants to take ownership over urban forest stewardship efforts, there are also a variety of ways the City can support this work. The community expressed a strong need for improving pathways for communication and collaboration between the City and the community. There are opportunities for improvement by increasing awareness and enforcement of the Tree Protection Ordinance and by promoting a clear and equitable tree removal permit application process. There are also areas where the City can promote tree plantings through redevelopment projects, promote planting climate-resilient species, and implement proper maintenance and management regimes for public trees. Each of these initiatives should be collaborative, considering how residents, community groups, and organizations can provide input on the updated processes and work in tandem to strengthen the health of the urban forest.

(above) Former Mayor Larry Moody's ceremonial tree planting in Jack Farrel Park on MLK Day 2017. (below) Park design charette flier. Photos: Canopy.

SECTION 3:
PLAN FOR THE FUTURE

IN THIS SECTION

6 STRATEGY and ACTION PLAN

Challenges, Benefits, and Recommended Steps

As a vibrant but socioeconomically disadvantaged community working to address systemic inequities in urban infrastructure and resources, East Palo Alto faces a number of challenges with regards to the urban forest. However, with proactive planning and care, these challenges can be overcome and lead the city toward the greener, healthier future imagined by the community. The many benefits provided by urban trees today will be compounded by future growth, if the urban forest can be continually supported through City policy, resources, and community action.

The following sections detail the recommended steps that East Palo Alto can take to reinforce its commitment to growing the urban forest. These recommendations are summarized in Table 6.1.

CHALLENGES FOR URBAN FOREST MANAGEMENT IN EAST PALO ALTO

Lack of Tree Canopy	Water Availability & Climate Change	Legacy of Urban Design	Unclear Authority & Responsibility
The overall tree canopy cover in the city is much lower than in surrounding communities. Both protection of existing trees, particularly large ones, and expansion of the urban forest are needed to increase the equitable distribution of this important type of urban infrastructure.	Portions of the city have high groundwater and are projected to see increased flooding and sea level rise with climate change, which will impact the potential for trees. Future drought and limited water for irrigation make low water use an important parameter for guiding tree species selection.	Existing urban design does not facilitate the growth of the urban forest, with few parks, small residential parcels, and narrow streets without space for tree planting.	Responsibilities for tree maintenance and permitting are not clearly defined in City code, reducing the City's ability to expand the urban tree canopy, to set consistent standards for tree care and maintenance, or to protect the existing tree canopy.

Reactive Approach to Tree Maintenance	Effectiveness of the Tree Protection Ordinance	Lack of Trust Between City & Community	Limited Resources
The City's current approach to tree maintenance responds to hazards and emergencies as they arrive, which can be inefficient, increase liability, and be detrimental for long-term tree health.	While the current tree protection ordinance provides some protection for existing trees, lack of enforcement, unclear requirements, and high fees are concerns for residents.	There is a lack of trust in the City's ability and willingness to provide the resources and maintenance needed to cultivate a healthy urban forest.	The current resources available for management of public trees limit the ability of the City to maintain a healthy urban forest.

Table 6.1. Actions and key points to meet the City's urban forestry objectives and goals.

Goal 1: Grow a healthy, extensive, vibrant, and diverse urban forest to provide 20% tree canopy cover by 2062 and 30% by 2122.		
Objective	**Action**	**Key Point**
1. Grow the urban forest.	1. Develop a city-wide tree planting strategy to increase equity within East Palo Alto and with neighboring cities.	1. Plant 200-250 public trees per year
		2. Plant 150-200 private trees per year
		3. Measure city-wide and zone-specific tree canopy cover every 10 years to track progress
	2. Plant climate-adapted species with the intention of planting the right tree in the right place to maximize benefits for public health, wildlife, and resilience.	1. Plant appropriate trees for each habitat zone in the city
		2. Plant the right species for the specific planting location
		3. Maintain species and age diversity within the urban forest
	3. Integrate trees into future urban design.	1. Set standard sizes for tree wells and planting strips to allow for healthy tree growth
		2. Enact a parking lot shade ordinance
		3. Review development plans for tree protection and planting
2. Define responsibilities and support improved maintenance practices and protections for public and private trees.	4. Codify responsibility and set standards for proactive public tree maintenance and protection.	1. Codify the City's authority and responsibility for street tree maintenance
		2. Take a proactive grid pruning approach to public tree maintenance
		3. Condense the City code regulating public trees into a single ordinance with all key components, including standards of care for public trees
	5. Revise the tree protection ordinance and implementation process to provide strong protection for private trees.	1. Revise the definition of a protected tree
		2. Revise the criteria for protected tree removal
		3. Require development projects to apply for tree removal permits
		4. Increase staff capacity to manage the tree protection ordinance by hiring or contracting with a certified arborist
		5. Review and clarify fees and requirements for tree removal permit applications
		6. Create a more robust appeals process with community input
		7. Publicize and enforce tree protections
	6. Seek opportunities for additional funding.	1. Identify opportunities to obtain additional funding to support urban forestry programs
Goal 2: Connect with an engaged and informed community to provide stewardship of the urban forest.		
Objective	**Action**	**Key Point**
3. Connect with the community around tree stewardship.	7. Design a cohesive and inclusive public outreach program focused on building awareness of the benefits of trees and how and why trees are protected in the city.	1. Engage with local community groups in urban forest stewardship activities
		2. Partner with Canopy to identify opportunities to expand current community outreach programs
		3. Provide up-to-date information about tree protections and management on the City website
		4. Translate tree information into common non-English languages spoken in East Palo Alto
		5. Designate a forum for the public to engage with tree management and protection actions
		6. Identify ways for the City to celebrate trees
	8. Become a Tree City USA.	1. As in Action 4, update the public tree ordinance to set standards of care
		2. As in Actions 5 and 7, designate or create a public body to review tree removal applications and appeals, and to serve as a public forum for tree-related issues
		3. As in Action 7, plan celebratory activities for Arbor Day

Goals

GOAL 1: Grow a healthy, extensive, vibrant, and diverse urban forest to provide 20% tree canopy cover by 2062 and 30% by 2122.

In order to achieve a more equitable distribution of tree canopy across the region and to secure the health, biodiversity, and environmental quality benefits of a healthy urban forest, the City of East Palo Alto should aim to grow the urban tree canopy. Trees should be planted strategically to address current inequities in the distribution of tree canopy cover in the city. In addition to increasing the overall tree canopy cover, this expansion should aim to maintain high species diversity, support a varied age structure, and select species that can provide the greatest benefit. Positive, proactive maintenance practices and protections can ensure the long-term health and vibrancy of the urban forest. This goal balances feasibility with the desired outcome of closing the "Green Gap" and reaching a similarly high tree canopy cover to what is seen in neighboring cities.

GOAL 2: Connect with an engaged and informed community to provide stewardship of the urban forest.

Successfully growing and stewarding the urban forest requires a strong partnership between the City, residents, involved organizations including Canopy, and developers. Community input showed that while many members of the community are eager to protect and plant trees, these partnerships are not yet solidified. Providing community members with accessible information about tree care and protections in the city can improve private tree stewardship. The City can provide leadership by effectively stewarding and championing public trees.

Objective 1: Grow the Urban Forest

ACTION 1. Develop a city-wide tree planting strategy to increase equity within East Palo Alto and with neighboring cities.

Key Points

- Plant 200-250 public trees per year
- Plant 150-200 private trees per year
- Measure city-wide and zone-specific tree canopy cover every 10 years to track progress

A strategic tree planting effort is necessary in order to reduce the disparity in tree canopy cover between East Palo Alto and surrounding cities. With the overall goal of achieving 20% tree canopy cover in the city by 2062, more trees will be needed particularly in places that currently have very low tree canopy cover and in places with more plantable area.

The amount of tree canopy cover created by a tree can vary widely depending on the tree species chosen and how good the conditions are where it is planted. Some tree species, such as crape myrtle, will remain small their entire life, making them suitable for locations with overhead utility conflicts or other space limitations, but less useful for creating tree canopy cover. Additionally, trees planted in small tree wells or strips without much soil volume will never grow as large as trees planted in large open areas like parks or yards. Existing trees will also grow and die over time, leading to further complexity in predicting how many new trees are needed to reach a tree canopy cover goal.

Assuming that the tree size distribution and planting conditions of future trees will match the existing public tree sizes (where, on average, tree canopies are about 19.6 ft in diameter), then we can estimate that about 15,100 new trees will need to be planted in the city to reach the goal of 20% tree canopy cover. If all planted trees were large trees in open conditions, as few as 2,000 new trees could be enough to meet the goal. However, if all of the new trees are small trees, it could take as many as 58,000 new plantings to reach 20% tree canopy cover.

Tree planting goals

Goal: 20% tree canopy cover by 2062

- **Plant 15,100 new trees**

- **Plant 350 to 400 new trees per year**

- **Plant 8,000 new trees along roads, in parks, and on other public lands**

- **Plant 2,800 new trees on residential land**

- **Plant 1,000 new trees on commercial and mixed use land**

- **Plant 3,300 new trees in the Ravenswood Business District /4 Corners Specific Plan area**

Creating a berm ensures water will irrigate this young tree's root ball. Photo: Canopy.

Based on current land use distribution in the city and the amount of pervious (plantable) area in each zone, we suggest tree canopy targets for each land use zone in order to add up to 20% tree canopy cover city-wide (Table 6.2). The highest targets are set for parks and schools, because the open space in these areas offer greater opportunities for planting healthy trees, and because the benefits of trees for park users and school children are particularly valuable. Residential parcels, mixed use areas, and roads are recommended to reach 20% tree canopy cover. These areas tend to have less open space for planting, but reaching higher levels of tree canopy can provide important benefits (see Chapter 3). Commercial areas have less opportunity currently as about 85% of these areas are paved or covered by buildings. Future changes to the way these spaces are designed will be important to further increase tree canopy (see Action 3). Finally, the Ravenswood Business District/4 Corners Specific Plan area currently has the lowest tree canopy cover in the city, but redevelopment offers an opportunity to expand the resource. The tree canopy cover target for this area is set at 20% to balance the opportunity of redevelopment with the feasibility of maintaining healthy trees along the edge of the marsh.

To meet these targets, the City should aim to plant about 200 trees per year in public areas (along roads, in parks, on school grounds, and on other public property) over the next 40 years. The City should also work to support planting of an additional 150-200 trees per year on private property and in the Ravenswood Business District/4 Corners Specific Plan area. The City can foster growth of the urban forest on private land by supporting tree plantings through local organizations and community groups (see Action 7).

Table 6.2. Tree canopy cover targets for 2062 by land use zone, set based on available pervious area to achieve 20% tree canopy cover city-wide.

Zone	Current canopy cover	Available pervious area	Target canopy cover	Approx. number of new trees needed*	Percent of all new trees needed**
Commercial	9%	15%	15%	300	2%
Mixed use	12%	22%	20%	700	5%
Residential (high density)	18%	22%	20%	200	1%
Residential (medium density)	18%	34%	20%	300	2%
Residential (low density)	17%	43%	20%	2300	15%
Ravenswood Business District/4 Corners Specific Plan area	6%	44%	20%	3300	22%
Parks and recreation zone	8%	86%	30%	2700	18%
Public institutional zone	9%	43%	30%	2000	13%
Roads and rights-of-way	13%	19%	20%	3300	22%
City total	**13.5%**	**37%**	**21%**	**15,100**	

* Assuming new planted trees match the size distribution of existing trees, reported to the nearest hundred trees.

** This percentage is the percent of all new trees planted that should be located in this zone in order to work toward tree canopy cover targets.

The City should monitor progress against these targets by evaluating the number of public trees in the public tree inventory and by periodically inspecting tree canopy cover across the city. An on-the-ground update to the existing public tree inventory should occur as soon as possible, in collaboration with the City's tree pruning contractor. In the future, the contractor should continuously update this inventory as services are performed (see Action 4). We recommend repeating a tree canopy cover analysis every 10 years to check whether tree canopy is expanding and the City is on track to meet the 40-year goal (Fig. 6.1).

Figure 6.1. Intermediate milestones for city-wide tree canopy cover. Regular updates to the tree canopy cover assessment should check whether these milestones have been achieved to determine if the city is on track to meet goals for 2062 and 2122.

TARGET TREE CANOPY COVER BY YEAR

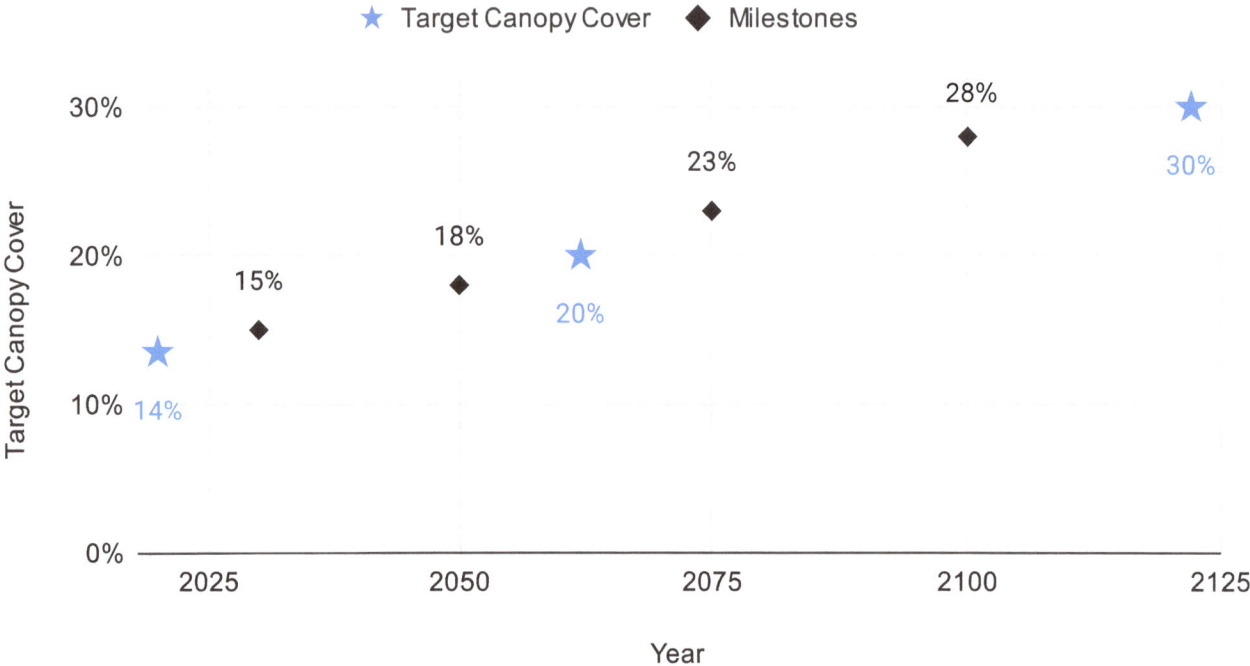

Residential streets of East Palo Alto. Imagery: Google Earth.

ACTION 2. Plant climate-adapted species with the intention of planting the right tree in the right place to maximize benefits for public health, wildlife, and resilience.

Key Points

- Plant appropriate trees for each habitat zone in the city
- Plant the right species for the specific planting location
- Maintain species and age diversity within the urban forest

Increasing the number of trees in the city's urban forest is important, but which kinds of trees make up the urban forest is also key to its long-term resilience and the benefits it can provide. Choosing the right tree for the right place is key for maintaining urban forest health over time and allowing trees to reach their full potential.

Planning and supporting a resilient urban forest

Diversity and Resilience

- Plant a diversity of tree species, with no more than 10% of any one species, 20% of any one genus, or 30% of any one family.
- Where possible, plant trees in parks and along city blocks at multiple times to create tree age diversity.
- Prioritize planting native tree species and appropriate climate-adapted species.

Human Health Impacts

- Plant trees with dense canopies along sunny active transit routes for the greatest cooling benefits.
- Plant diverse species to avoid synchronous pollen production.
- Plant trees in high density and high traffic areas to increase people's experience of and interactions with trees.

Native Landscapes and Wildlife

- Plant native trees, especially oaks, to support a diversity of wildlife and connect with the local natural history.
- Prioritize planting in parks to create habitat patches and along corridors to build connected networks of greenspace for wildlife.
- Avoid tree planting along the tidal marsh, which could negatively impact endangered marsh animals by increasing predation.

Water Use

- Prioritize low water use species.
- Support young trees with supplemental irrigation if needed as they become established.
- During times of severe water limitations, prioritize irrigating larger, older trees that provide irreplaceable ecosystem services.
- Plant salt-tolerant species if recycled water will be used.

Fruit Production

- Plant fruit trees in private yards and public spaces where fallen fruit will not create hazards for pedestrians or vehicles.
- Balance fruit tree planting with planting native and climate resilient trees that can provide a wider range of ecosystem services.

Incorporating a diversity of trees to address these needs is an important goal for the urban forest. Particular drivers should be prioritized where they are most important in the city. For example, air pollution mitigation is key near the US 101 Highway, heat mitigation is especially important for playgrounds and active transit corridors, and protecting wildlife is a priority near the Baylands and around parks that can serve as habitat patches. The best locations for each type of tree also depend on the local conditions: some trees are better able to tolerate saltwater intrusion, which might occur near the Baylands, while some are more likely to thrive in well-drained upland soil types. Small tree wells might only be able to accommodate smaller-sized trees, which provide fewer benefits but are better than having no tree at all. Meanwhile, big open planting areas in parks or on public grounds are great locations for larger trees that need more space, and can provide exceptional habitat, beauty, and shade.

Different trees can be prioritized in parts of the city with different needs and constraints. Tree planting zones in East Palo Alto were developed by considering historical, present, and projected future conditions in East Palo Alto, including the city's native vegetation types, hydrology, soils, and climate (Fig. 6.2). There are three core objectives driving the planting zone map: adapting to sea level rise, mitigating impacts of the US 101 Highway, and supporting local native biodiversity. In addition, priority areas highlight locations where people spend time outdoors: in parks, in schoolyards, and along bike routes. In all zones, trees planted should be drought-tolerant to suit local climate conditions. Habitat zones can guide tree choice broadly across the city, but specific site conditions are crucial for determining a tree's future success. Table 6.3 describes each of these habitat zones and offers a sample list of native trees that are generally suitable in each zone. A qualified arborist can help further identify tree species that are suitable for sites with different types of constraints.

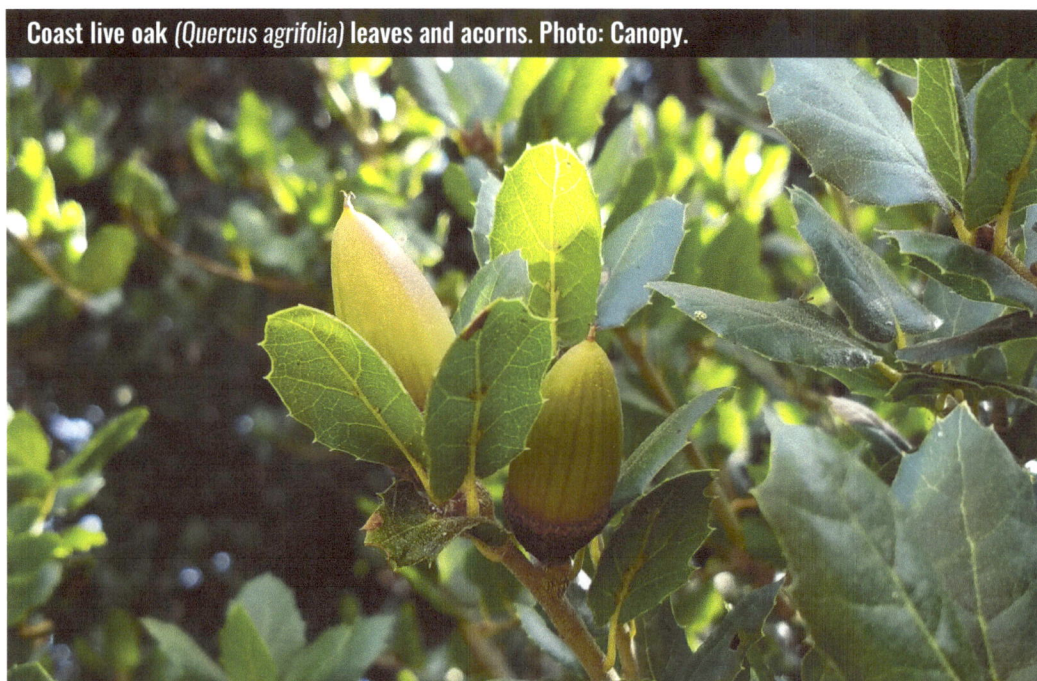

Coast live oak *(Quercus agrifolia)* leaves and acorns. Photo: Canopy.

Habitat Zones
- US 101 Highway Corridor
- Oak woodland habitat zone
- San Francisquito Creek riparian zone
- Tolerant of high water table
- Salt tolerant
- Baylands

Priority Areas
Active Transit Corridors
- Bicycle path
- Bicycle lane
- Bicycle route
- Park
- School

SAN FRANCISCO BAY

0.5 miles

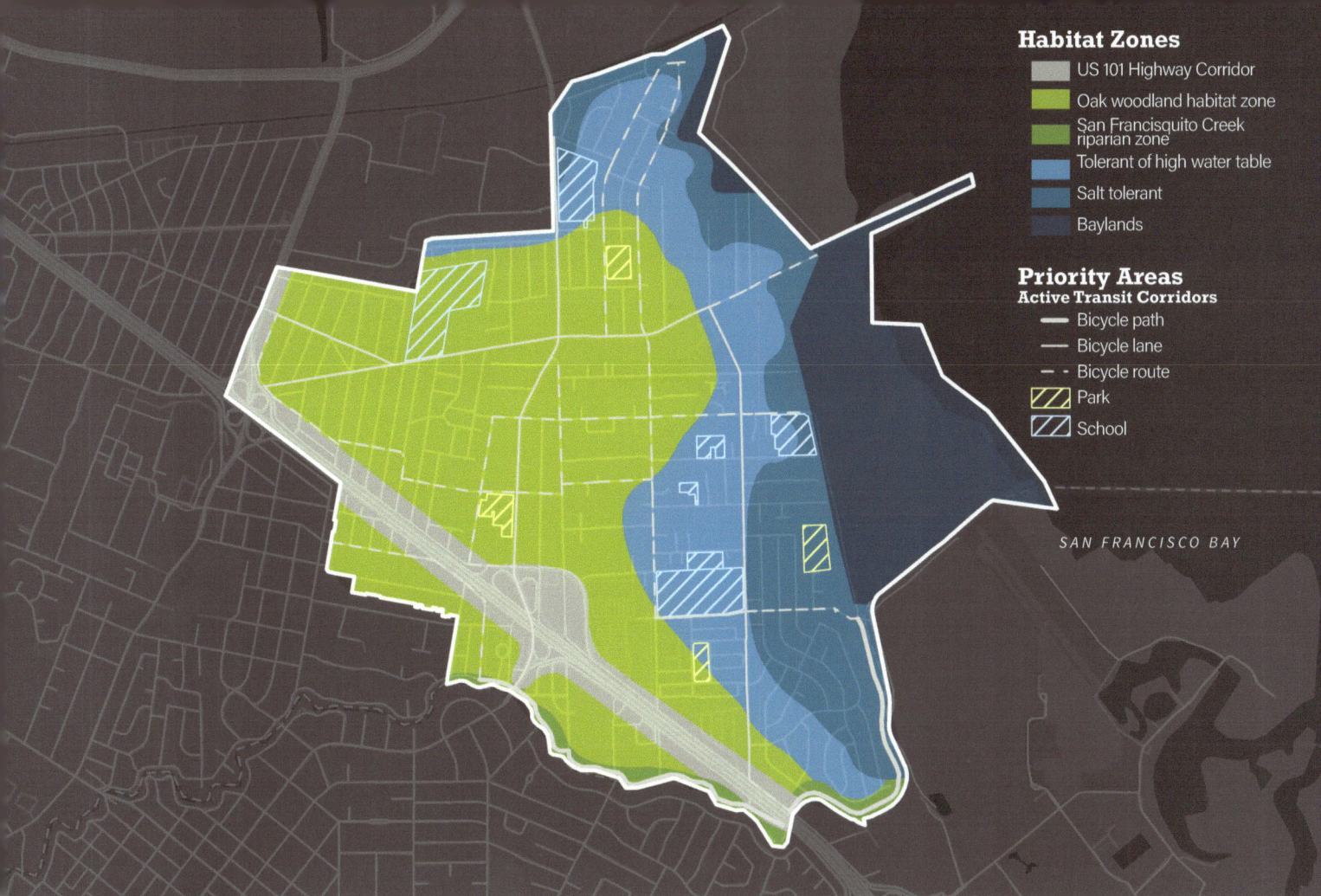

Figure 6.2. Map of habitat zones and special priority areas for strategic tree planting.

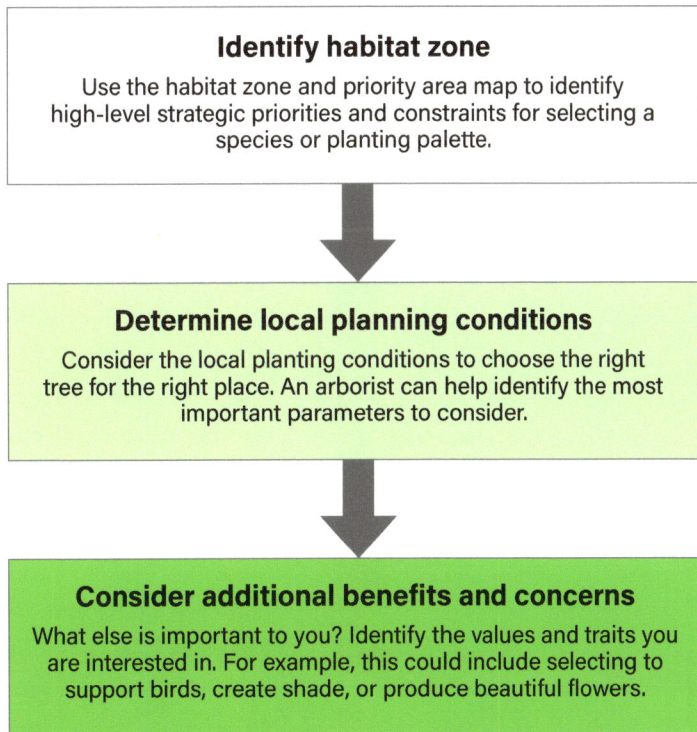

Identify habitat zone
Use the habitat zone and priority area map to identify high-level strategic priorities and constraints for selecting a species or planting palette.

Determine local planning conditions
Consider the local planting conditions to choose the right tree for the right place. An arborist can help identify the most important parameters to consider.

Consider additional benefits and concerns
What else is important to you? Identify the values and traits you are interested in. For example, this could include selecting to support birds, create shade, or produce beautiful flowers.

Table 6.3. Descriptions of broad habitat zones in East Palo Alto, each of which was developed based on the zone's unique physical conditions, objectives, and opportunities. Examples of recommended native species suitable for these habitat zones are also included, although this list is not comprehensive.

Zone	Description	Example Recommended Native Species
HABITAT ZONES		
US 101 Highway corridor zone	Along the US 101 Highway-corridor, trees provide important services, acting as a sound wall and filtering air pollution. Surrounding the US 101 Highway-corridor, the tree planting palette should prioritize species with the potential to capture pollutants and insulate sound, while also having low emissions of volatile organic compounds (VOCs). In addition, these trees should be able to tolerate vehicular pollutants such as ozone, nitrogen dioxide, and sulfur dioxide. Where planting areas are more spacious along this corridor, large trees should be selected.	• Bigleaf maple (*Acer macrophyllum*) • Pacific madrone (*Arbutus menziesii*) • California ash (*Fraxinus dipetala*) • Oregon ash (*Fraxinus latifolia*) • Gray pine (*Pinus sabiniana*) • Hollyleaf cherry (*Prunus ilicifolia*) • Coast redwood (*Sequoia sempervirens*) • California bay laurel (*Umbellularia californica*)
Oak woodland zone	Before the city's development, 40% of the city's landscape was dominated by oak woodlands and savannah. The city can support a highly functioning urban forest by prioritizing native trees, especially oaks, in more upland locations where they once thrived.	• Coast live oak (*Quercus agrifolia*) • Canyon live oak (*Quercus chrysolepis*) • Blue oak (*Quercus douglasii*) • Oregon white oak (*Quercus garryana*) • California black oak (*Quercus kelloggii*) • Valley oak (*Quercus lobata*) • Interior live oak (*Quercus wislizeni*) • Pacific madrone (*Arbutus menziesii*) • California buckeye (*Aesculus californica*) • Hollyleaf cherry (*Prunus ilicifolia*)
San Francisquito Creek riparian zone	The San Francisquito Creek is an important corridor for both people and wildlife. This zone can support riparian trees that are naturally found along waterways, and that provide ample shade for recreation and active transit, improve the health of the creek, and host local biodiversity.	• Bigleaf maple (*Acer macrophyllum*) • California boxelder (*Acer negundo* var. *californicum*) • White alder (*Alnus rhombifolia*) • Oregon ash (*Fraxinus latifolia*) • Pacific wax myrtle (*Morella california*) • Western sycamore (*Platanus racemosa*) • Black cottonwood (*Populus balsamifera* ssp. *trichocarpa*) • Fremont cottonwood (*Populus fremontii* ssp. *fremontii*) • Coast live oak (*Quercus agrifolia*) • Valley oak (*Quercus lobata*) • Red willow (*Salix laevigata*) • Arroyo willow (*Salix lasiolepis*) • Yellow willow (*Salix lucida* ssp. *lasiandra*) • California bay laurel (*Umbellularia californica*)

Zone	Description	Example Recommended Native Species
HABITAT ZONES		
Salt tolerant zone	Planting salt-tolerant and high-groundwater-tolerant trees in the zone most immediately threatened by sea level rise and related saltwater intrusion will support the growing urban forest's resilience to a changing climate. Tree planting should be avoided in and directly adjacent to tidal marsh areas, where no trees were historically found.	• Western sycamore (*Platanus racemosa*) • Pacific wax myrtle (*Morella californica*) • California boxelder (*Acer negundo* var. *californicum*)
Tolerant of high water table zone	The region beyond the immediate shoreline will also be impacted by sea level rise—at first, by rising groundwater levels, and in the long-term, by increased flooding and inundation. Trees tolerant of a high water table will be best adapted to rising groundwater levels.	• Bigleaf maple (*Acer macrophyllum*) • California boxelder (*Acer negundo* var. *californicum*) • White alder (*Alnus rhombifolia*) • Brown dogwood (*Cornus glabrata*) • Creek dogwood (*Cornus sericea* ssp. *sericea*) • Oregon ash (*Fraxinus latifolia*) • Pacific wax myrtle (*Morella californica*) • Western sycamore (*Platanus racemosa*) • Black cottonwood (*Populus balsamifera* ssp. *trichocarpa*) • Fremont cottonwood (*Populus fremontii* ssp. *fremontii*) • Red willow (*Salix laevigata*) • Arroyo willow (*Salix lasiolepis*) • Yellow willow (*Salix lucida* ssp. *lasiandra*) • California bay laurel (*Umbellularia californica*)
PRIORITY AREAS		
Active transit corridors	Active transit corridors, including current and planned bicycle lanes, routes, and paths, are important areas for prioritizing street tree plantings. Properly maintained, large shade trees benefit pedestrians and bicycles by cooling, improving roadside aesthetics, calming traffic, and creating more appealing routes for biking and walking. Forested streets also create corridors for wildlife to move through the city, increasing connectivity.	
Park	Parks are important opportunity areas for tree plantings. Trees in parks promote physical and mental health benefits for park visitors, create shade and cool down nearby neighborhoods, and provide habitat to support local biodiversity. The city can maximize benefits from park trees by planting large, native trees that provide plentiful shade, aesthetic value, and habitat. In addition, prioritizing tree planting around neighborhood parks can increase the effective park size by creating more habitat for wildlife.	
School	Trees provide a diversity of benefits for schools, not only by generating shade and mental and physical health benefits, but also by offering opportunities for experiential learning, nutrition, and social connections. Planting, maintaining, and harvesting from fruit trees creates opportunities to learn, builds community outdoors, and fosters a connection with healthy foods. Large, native shade trees also create cooler spaces for outdoor recreation and opportunities to learn about and connect with local plants and wildlife.	

ACTION 3. Integrate trees into future urban design.

Key Points

- Set standard sizes for tree wells and planting strips to allow for healthy tree growth

- Enact a parking lot shade ordinance

- Review development plans for tree protection and planting

Redevelopment and construction projects present important opportunities to design trees into the urban fabric. Projects should aim to maximize tree planting in open space areas and along streets to gain the benefits of tree canopy in these newly built areas. In addition to protecting the most valuable existing trees, projects should design sufficient space to support healthy new trees in order to achieve long-term tree canopy growth. Review of development plans for tree planting sites is one way to support this goal, and adding shading requirements to City tree ordinances is another. Embedding tree planting into future urban design is essential to enable East Palo Alto to expand the urban tree canopy and reach the 100 year goal of 30% tree canopy cover.

Design roadways and building sites that incorporate tree planting areas

During redevelopment projects, incorporating space for trees along the street and within the site is crucial to achieving tree canopy cover goals. Incorporating large planting areas helps streets and private landscapes support more tree canopy. Trees are more likely to survive in larger planting areas, making planting in these areas a better long-term investment (Lu et al., 2010). In addition, trees can reach a greater size in larger planting areas, generating more tree canopy cover (Sanders & Grabosky, 2014; North et al., 2018; Dahlhausen et al., 2016). Street and private development design choices such as including structural soils, pervious pavement, and larger planting spaces can prevent conflict between trees and infrastructure and help promote healthy trees (Mullaney et al., 2015; Berland et al., 2017).

The City could set standards for planting areas, like making a 4x6 foot planting strip or 4x6 foot tree well an absolute minimum. Soil depth is important too, and trees should be provided with soil that is at least 3 feet deep, and ideally 4 or 5 feet deep. The US Environmental Protection Agency recommends a minimum of 500 ft^3 per tree and suggests 600 ft^3 for small trees, 1,000 ft^3 for medium trees, and 1,500 ft^3 for large trees (U.S. Environmental Protection Agency, 2016).

Require shading over parking lots

In the municipal code, Chapter 13.24.430 (Parking areas) currently defines landscaping requirements for parking areas, and Chapter 18.28.030 (Landscape requirements) sets development standards for tree planting. Enhancing either of these sections to require tree planting in parking lots would improve tree canopy cover in areas where the shade is most needed. Several cities in the Central Valley, where summer temperatures are often over 100°F, have added shade ordinances or shade components to cool these spaces. In addition to reducing heat and solar radiation for cars and pedestrians, studies have found that tree shade extends the life of pavement by protecting it from solar damage (McPherson & Muchnick, 2005), and keeping parked vehicles cooler reduces the passive emission of air pollutants caused by gasoline evaporation (Scott et al., 1999).

A shade requirement could consist of a certain percentage of parking lot area covered by tree canopy by 15 years after construction or adoption of the ordinance. Both Davis and Sacramento require 50% of parking lots to be covered within 15 years. This approach also aligns with the California Green Building Code Section A4.106.7, which requires 50% shade coverage over parking lots within 15 years, and 20% shade over landscape and hardscape such as driveways and walkways in landscape areas (International Code Council, 2019).

To add this provision to City code, the City would need to:

- Set a tree canopy cover target for parking lots

- Define which parking lots would be required to comply

- Define the surface areas to be included in the calculation (e.g., parking stalls, drive aisles, maneuvering areas)

- Indicate if public trees or trees on neighboring properties could be included in the calculation

- Define a process to review plans for compliance

The City could support developers by producing shade tree guidelines to simplify the process for calculating parking lot shade, as has been done by the City of Pleasanton (City of Pleasanton, 2020). The guidelines could provide a list of approved trees and their canopy sizes, recommended tree planting practices that encourage the use of structural soil systems such as Silva cells, and guidance for choosing an appropriately diverse mix of trees.

Set clear processes and goals for development project review

Redevelopment is an important opportunity to expand tree canopy cover over time. As redevelopment plans are approved by the City, staff should develop regular procedures to confirm tree protection and preservation are carried out, and tree removals should be approved following the standard process (see Action 5).

In addition to adding and reviewing for tree well sizing and parking lot shading as described above, the City should set ambitious tree canopy cover targets or planting requirements for development projects, following from the targets outlined in Action 1. Development plans should be reviewed by a certified arborist to ensure that tree protections during construction are sufficient and the site has been designed to support healthy trees.

Objective 2: Clarify and codify responsibilities, maintenance practices, and protections for public and private trees.

ACTION 4. Improve public tree maintenance and protection

Key Points

- Codify the City's authority and responsibility for street tree maintenance.

- Take a proactive grid pruning approach to public tree maintenance.

- Condense the City code regulating public trees into a single ordinance with all key components, including standards of care for public trees.

The City code that governs public trees, Chapters 12.16 and 13.24.410, does not define who is responsible for the maintenance of public trees, what condition City trees are to be maintained to, how trees are to be planted, when trees are to be removed, who will maintain the inventory of public trees, who will maintain the street trees on each block, and who will maintain a budget for such activities. City staff are currently maintaining trees with a "reactive" maintenance approach: trees are pruned and removed on an as-needed basis, which often results in poorer tree health, more emergency tree removals, and increased spending for tree care.

This recommended action contains two elements: clarify authority for public tree maintenance and move the City to a proactive maintenance approach. Success of these actions hinges on both parts being implemented together and reflected in updated City code. Designating the City's responsibility for public tree maintenance without moving to a proactive maintenance program will leave the City with liability beyond what can be handled with current staffing and processes. Moving to a proactive maintenance program without defining staff's roles and responsibilities will not give staff the authority to complete the necessary work.

Clarify authority for street trees

The City's public tree ordinance should state who is responsible for the care and maintenance of trees in the right-of-way. It is generally understood that trees in the area between the curb and the sidewalk (typically called street trees, parkway trees, or trees in the public right-of-way) are considered City trees. In most cities, the responsibility for maintenance and care of street trees is defined by City code. Defining the roles of the City and adjacent property owners can help avoid misunderstandings and can alleviate the City of liability.

In general, cities allocate the responsibility for street tree maintenance with two different strategies. One strategy has adjacent property owners maintain

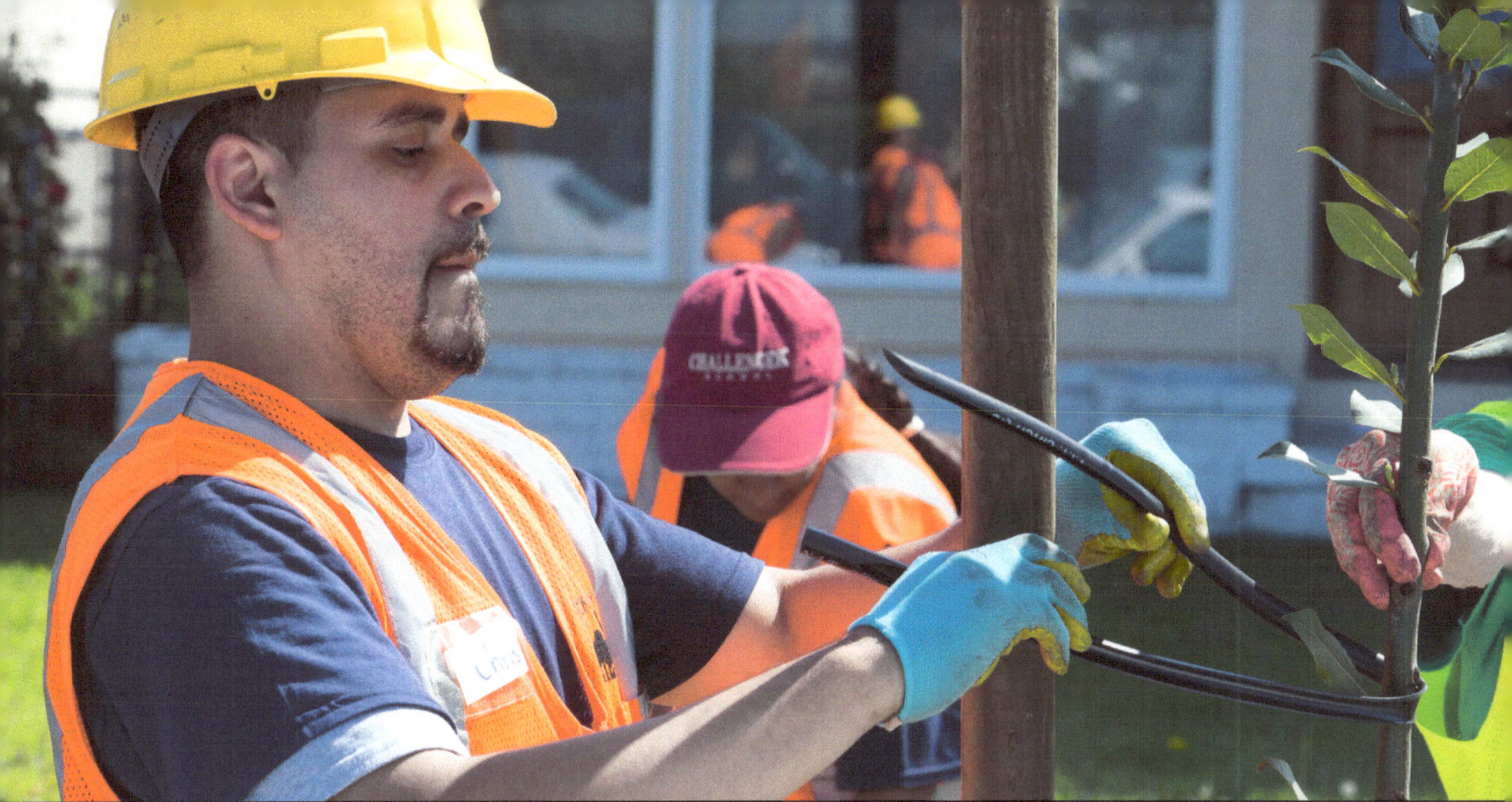

Staking provides stability to young trees. Photo: Canopy.

the street tree in front of their home. Using this strategy, the property owner can choose an arborist to assist with maintenance and care of the street tree. The property owner also takes on the cost of any necessary tree maintenance, releasing the City from fiscal responsibility at the expense of the property owner. In this scenario, the municipality does not control tree care but does maintain liability, and lack of care can lead to failures for which the City can be held liable.

This strategy is used in San Jose and Vallejo. In San Jose, the City has five municipal inspectors who ensure that maintenance of trees in the right-of-way is performed in accordance with City standards. Code enforcement staff are responsible for citing anyone who fails to comply with the City standards. In Vallejo and San Jose, property owner-maintained trees are also supported by City crews or contract crews who prune street trees within Landscape Maintenance Districts. Landscape Maintenance Districts are special districts formed to provide benefitting property owners the opportunity to pay for enhanced landscaping improvements, maintenance, and services beyond those generally provided by the City. This model can work with an investment in public oversight, but can also lead to disparities among different areas of a city.

The second strategy is for the City to maintain trees in the right-of-way, which is a proven tactic for both maintaining and growing the urban forest through proprietary oversight. Nearby cities such as Palo Alto and Menlo Park use this approach, which also is currently performed in East Palo Alto, though authority is not set by the City code. With this approach, the City directly invests in the maintenance of street trees, which requires staffing and budget. The City gains the insurance that trees are well maintained by skilled, certified professionals, and are poised to take action to limit tree-related liability. Taking this approach will best enable East Palo Alto to appropriately maintain public trees.

Transition to a proactive cycle pruning program

With the City responsible for public tree management, maintenance can be either proactive or reactive. A reactive maintenance program is a program that responds to tree failure, complaints, and hazards, as opposed to a proactive program with a regular inspection cycle that identifies trees in need of care before they become a hazard, fail, or are a public nuisance. In reactive maintenance programs, trees are left to grow and fail without intervention from the tree managers. This leaves the tree manager, in this case the City, open to liability. In the eyes of the law, lack of knowledge of a hazard does not relieve or alleviate the tree owner of responsibility to address the hazard.

We recommend that the City transition to a proactive approach, in which public trees are inspected and pruned on a regular cycle ("grid pruning"). Maintenance costs in a proactive program are often less onerous once trees are being regularly maintained. This approach also reduces the risk of tree failure and associated liability. A major added value is that the community sees the City taking positive action for trees.

Per-tree pruning costs in East Palo Alto have been high in comparison to surrounding cities in part because of deferred maintenance. In a literature review, researchers showed that overall, the costs of not maintaining urban trees were greater than the cost of maintenance applied at the right time (Vogt et al., 2015). Small trees can be pruned inexpensively to fix structural issues, which become much more costly, difficult, and potentially harmful to the tree's health when left until the tree is large. With a regular pruning cycle, structural issues can be identified and resolved when the tree is young. Trees that are properly pruned when young will be more successful street trees for the long term than trees that are left with structural problems.

Information gathered from other municipalities (including Vallejo, Newark, Santa Clara, Holister, Gilroy, and Palo Alto) indicates that with grid pruning in place, the cost to prune any tree would be about $140, as opposed to the $280 per tree on average the City is paying now. Taking a grid pruning approach would move the City from pruning about 300 trees annually to 700 trees, and for half the cost per tree.

Effective urban forest management requires the ability to prioritize work, document activity, and map the current tree inventory for strategic planning. Access to the current composition of the urban forest is a critical element in achieving these goals. Best practice requires that East Palo Alto maintain an up-to-date tree inventory that tracks these items. There are a variety of dynamic tree management software systems available for this purpose which are designed to create work orders and reports, develop maintenance schedules, store work histories and related documents for each site, create GIS overlays with sewer lines and other potential infrastructure conflicts, generate vacant planting sites, and calculate tree canopy cover, tree valuation and benefits. Tree maintenance contractors typically have software with these capabilities available, and future contracts could include management and upkeep of the tree inventory. Alternatively, City staff can use any non-dynamic program (including a spreadsheet of any kind) to manage data. The goal is for the tree manager to easily access information on tree species, condition, location, and maintenance history.

To move to a proactive maintenance program, the City will need to:

- Update the public tree ordinance to set the maintenance standard.

  - Set the goals of the program and policies for maintenance that describe who is responsible for what and where the funding for the program will come from.

  - Draft a standard of care statement where the appropriate and acceptable care for trees is defined.

- Define record keeping protocols to ensure that the same information is collected every time maintenance occurs.

  - The contractor hired to conduct grid pruning can also update tree inventory information as part of the scope of work, and provide the City with an up-to-date inventory of all public trees to reference.

- Assess and prioritize. Put all public trees on a grid pruning schedule, prioritizing streets with the most pedestrian traffic or in downtown areas.

- Set the budget and pruning cycle. Currently, $120,000 is allocated annually for tree pruning and removal contract services. Based on the tree inventory and the current pruning costs for grid pruning, the project team estimates that moving to a 7-year grid pruning cycle would require an annual budget allocation of $120,000 for contracted pruning and an additional $30,000 for emergencies, out-of-cycle pruning, and tree removals. These estimates do not include East Palo Alto staff time.

Revise the City code governing public trees

The two ordinances (Chapters 12.16 and 13.24.410) that currently govern public trees should be revised into a single public tree ordinance (Chapter 12.16) with all necessary components (see box text).

The public tree ordinance should reference City tree maintenance guidelines and strategies, setting a standard for City staff to follow, as described previously. It should also set criteria and processes for removal of public trees. These criteria and processes may be the same as for the removal of protected private trees, or may be different (see Action 5). The public tree ordinance should also set replacement requirements for when public trees are removed. Replacement requirements could be planting 2 or 3 trees for each tree removed in order to grow the urban tree canopy over time.

In addition to requiring replacement plantings for public trees that are removed, the City could also create a fund specifically for street tree maintenance activities, to which the City could contribute if a situation arises where a replacement tree cannot be planted. Covered activities could include planting, watering, small tree care, and pruning, but not removals or park maintenance activities unless specifically related to tree care. This fund can help to offset costs of a more robust program. In-lieu fees should be set considering both the cost of planting and young tree maintenance.

Finally, the public tree ordinance should set planting standards and processes. If the City wishes to maintain a street tree management plan identifying species to be planted on individual streets, then it will need to delineate a process for residents to request changes to the plan. Tree planting and maintenance standards and specifications should be in accordance with American National Standards Institute (ANSI) A300 standards. If East Palo Alto ties the City standards for tree care to the ANSI standards, the City standards will remain up to date as changes to best practices arise without further action by the City.

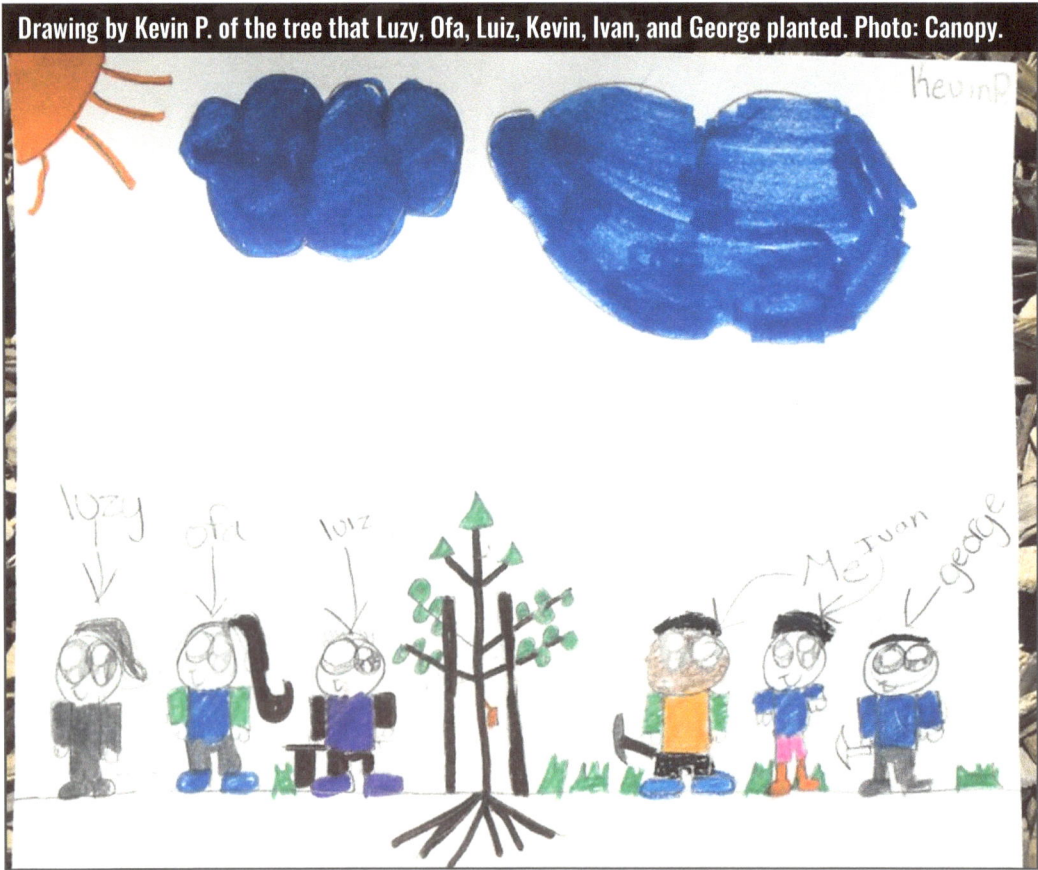

Drawing by Kevin P. of the tree that Luzy, Ofa, Luiz, Kevin, Ivan, and George planted. Photo: Canopy.

Components of a street tree ordinance

A meaningful street tree ordinance includes the following:

- **A title** - A clear concise title should set the stage for the document

- **The intent and purpose of the code** - This section needs to outline the values that led the City to codify this provision

- **Definition of terms** - Terms like "tree," "City tree," and "public right-of-way" need to be defined in this section and consistent with terms in the tree protection ordinance

- **Authority** - This section defines who is responsible for what actions under what circumstances

- **Policies and procedures** - In this section the City has the opportunity to set the work plan and standards of care for the urban forestry team. The team will be accountable to the City Council for these actions being accomplished as laid out. Additionally, this section should include:

    - Noticing requirements for pruning, planting, and removal

    - Applications process for removal of trees on public property and appeals process

    - Tree replacement noticing, planting standards, and process for requests not to have a tree planted

    - Applications for pruning on public property, standards for pruning, noticing in advance of pruning, appeals process for denied tree pruning applications

Trees along US 101 Highway remove harmful pollution from the air. Photo: Canopy.

ACTION 5. Revise the tree protection ordinance and implementation process to provide strong protection for private trees.

Key Points

- Revise the definition of a protected tree
- Revise the criteria for protected tree removal
- Require development projects to apply for tree removal permits
- Increase staff capacity to manage the tree protection ordinance by hiring or contracting with a certified arborist
- Review and clarify fees and requirements for tree removal permit applications
- Create a more robust appeals process with community input
- Publicize and enforce tree protections

The existing private tree protections within Chapter 18.28 of the City's Development Code could be made more effective through improvements in two areas:

1. Revising the ordinance to update tree removal criteria, clarify requirements for development, and add new measures to support increased tree canopy cover; and

2. Standardizing the processes associated with tree removal permit application, review, and enforcement.

Ordinance revisions

As with the public tree ordinance, the private tree protection ordinance should include a clearly stated goal, authority and responsibilities, and policies and procedures. Terms should be consistently defined in both the public and private tree ordinances. The purview of this ordinance would be to protect trees during the development process, determine appropriate mitigation for trees removed, and enforce criteria for private tree removal permits. Street trees and trees in public places would be under the jurisdiction of the Public Works Department as defined by the public tree ordinances (Action 4).

Revise the definition of a protected tree

Under the current ordinance, a "protected tree" includes all of the following:

A. Any tree having a main stem or trunk that measures 24 inches or greater in circumference at a height of 40 inches above grade;

B. Tree within a public street or public right-of-way, regardless of size;

C. Any tree that was required to be preserved as a condition of a development approval granted by the City;

D. Any tree required to be planted as a condition of a development approval granted by the City; and

E. Any tree required to be planted as a replacement for an unlawfully removed tree.

In the handout used by the City to inform the public of tree removal permit requirements, Protection A is given as, "Any tree with a main stem or trunk that measures forty (40) inches or greater in circumference at a height of twenty-four (24) inches or two (2) feet above natural grade." This is the definition currently applied by City staff resulting in fewer trees protected in practice than by the ordinance text.

These classes of protection require revisions to match best practices and the specific needs of East Palo Alto.

Tree ordinances often use size thresholds to define which trees should be protected. These thresholds are typically based on tree diameter measured at a standard height of 54 inches above grade. Protection A should be revised to use this metric. We recommend protecting trees over 8 inches in diameter at this standard height, which will be similar to the current ordinance protection of trees 24 inches in circumference (7.6 inches diameter).

If this ordinance is to focus on private trees while the ordinance in Chapter 12 pertains to public trees, the protection classes should be revised to remove reference to public trees (i.e., remove Protection B and specify "any *private* tree" in other classes).

Protections C, D, and E show the City's interest in protecting trees planted as replacements or as part of development agreements. In practice, protecting these trees requires staff to track all trees that meet these conditions and to enforce rules pertaining to their removal. With current staffing and procedures, it is very difficult for the City to adequately enforce these protections. The City could consider removing these protection criteria if they cannot be enforced, or should plan to allocate staff time toward tracking these trees in the tree inventory and enforcing penalties for unapproved removals.

Revise the criteria for removal

Having tree removal criteria that reflect the values of tree preservation and growth of the urban forest, yet are flexible enough that applicants can navigate the implementation without intervention, is crucial. The criteria should be simple and clear, such that there is little room for conflict. Currently, protected trees may be removed if they meet one of three criteria:

1. The tree is dead or infected with a terminal disease.

2. The tree is structurally unsound and cannot be corrected or the risk cannot be significantly reduced by traditional pruning, cabling, or bracing.

3. The tree is causing visible damage to property, which cannot be corrected without destroying the tree canopy or root system.

The current criteria do not clearly address removals associated with development or with removals by governmental agencies, leaving these removals outside the scope of the tree protection ordinance. Review is included in the total project review.

An added criterion related to development should seek to allow removal of protected trees only in circumstances where protected trees would stop the applicant from developing the property, which would be considered under the law as a "taking of real property." Healthy protected trees should be preserved in all other circumstances. The goal of this criterion would be to remove ambiguity around the requirements for an allowable removal during development, and to require a high level of evidence that no feasible alternatives to removal exist.

Additionally, a criterion should be added to allow for the removal of protected trees that interfere with an existing or planned public utility, transportation facility, wildlife habitat restoration project, or other governmental agency project. This criterion should be written such that the onus is on the applicant agency to provide proof of the tree's interference with existing or planned public infrastructure and to show that preservation of the tree would add unreasonable cost to the project.

Remove the exemption for development projects

Currently, no tree removal permit is required if tree removal has been authorized as part of a development approval granted by the City. However, to adequately provide protections, public notice and documentation of tree removals, and an appeals process during development, development projects should be subject to at least the same level of requirements as non-development removals. We recommend that development projects be required to submit a tree removal permit application for trees removed in addition to a development permit application. Many other cities, including Menlo Park, Palo Alto, Hayward, and Berkeley, require separate permits for trees removed with development projects.

Expand the appeals process to include community voices

Based on feedback received from other communities and from the residents of East Palo Alto, the existing public appeals process should be enhanced to allow members of the community who are not satisfied with the result of their application or neighbors who are unhappy with the removal of a protected tree to express their concerns and understand the reasoning behind permit approvals or denials. We recommend that updates of approved tree removals be provided to a public body who would hold the authority to hear appeals. This body can be an existing public body or a newly created public body. Both the authority and responsibilities of this body should be clearly stated and incorporated into the private tree protection ordinance. Costs for bringing a case to this body should be limited to allow equal access by all members of the community.

Review of permit applications and hearing appeals would require staff time to prepare as well as time on agendas of the designated public body. However, it would provide value by increasing transparency around tree removal permit decisions and allowing the public to have a voice in these decisions. As an interim step, transparency around tree removal permitting decisions could be increased by regularly presenting a status update on tree removals in the city at an existing public meeting. While not increasing public participation, this step would allow the community to see how trees are removed or protected.

Clarify requirements for arborist reports

As a requirement for the removal of private trees, tree owners must currently obtain a report from a certified arborist describing the general characteristics and health of the tree to be removed to ensure that the tree meets the criteria for allowable removal. This can be an important step to prevent removals of trees that are mistakenly believed to be dead or in poor health. Since 2018, the City has required applicants to select an arborist from a list of certified arborists maintained by the City (Section 18.28.040). This restricts tree owners' ability to choose an arborist that best fits their budget and needs, and requires the City to maintain an up-to-date list of contractors. In response to community concerns, City staff began accepting arborist reports written by any ISA certified arborist, which is a sufficient and less burdensome requirement. The ordinance should be updated to reflect this practice.

Tree protection processes

Hire, train, or contract with a certified arborist

Municipalities most effective in administering urban forest master plans across city departments have a credentialed arborist on staff, who can be responsible for all tree-related activities and have the authority to carry them out. Alternatively, a qualified outside consultant can provide support for necessary functions.

A certified arborist should be responsible for reviewing all tree removal permits, including those for redevelopment projects, and all tree protection plans for construction. Currently, 15 to 25 permit applications are submitted annually prior to tree removal. An additional 20 or so applications per year are submitted for trees that have already been removed, which cannot be evaluated with the same level of detail. An unknown number of additional removals are associated with development, which do not currently go through the same tree removal permit application process. An arborist will have the skills and knowledge to determine whether a tree is a good candidate for protection or when removal should be permitted. Additionally, a certified arborist can contribute to development and oversight of public tree management plans and City tree planting efforts.

Consider reducing fees to increase compliance

Based on discussion with permit applicants, we found that the costs of complying with the tree removal permitting process were onerous and likely to result in non-compliance. If the cost of a removal permit can be lowered in cases where the non-development criteria are met (i.e., the tree is dead, dying, or hazardous), it would likely encourage applicants to follow the process and plant more trees. Review of protected tree removal applications being performed by a certified arborist would likely decrease staff time spent. Development-related permit fees could be increased to cover costs for the condition-based removal permit fees.

Enforce tree protections

Enforcement of tree protection policies is crucial to the success of the ordinance. Many cities struggle with how best to enforce a tree preservation ordinance. These struggles can arise from a lack of public education on how to follow the policies, processes that are difficult for applicants or staff to follow, or when enforcement capacity is lacking.

Pleasanton provides an example of a successful, comprehensive enforcement strategy. Information posted online is accurate, clear, and easy to follow. Staffing of the urban forestry group meets the needs of the City, and code enforcement assists with writing citations which lead to penalties. Code enforcement citations have led to violators being charged the maximum penalty ($5,000), which has had the impact of dissuading future violations.

Currently, East Palo Alto code enforcement staff write citations for violations to the tree ordinance only if they are reported by community members. Shifting code enforcement staffing such that some hours are spent checking for violations over the weekend would help identify and reduce illegal tree removals on the weekend. Additionally, we recommend that there be a dedicated line and email address for community members to call or email to report a violation. Increased enforcement should be coupled with a robust community information program and enhanced educational materials to ensure that community members are aware of the protections in place and the penalties for unpermitted removal of a protected tree (see Action 7).

(left) Former Mayor Regina Wallace Jones and her ceremonial tree MLK Day 2020. (right) Former mayor Carlos Romero planted a Silver Linden *(Tilia tomentosa)* on MLK Day 2021. Photos: Canopy.

Action 6. Seek opportunities for additional funding

Key Points
- Identify opportunities to obtain additional funding to support urban forestry programs

Canopy has partnered with the City to seek and raise funding for forestry-related projects in East Palo Alto, including a tree inventory grant, numerous tree planting grants, and funding for the development of this Urban Forest Master Plan, all from Cal Fire. The City could seek additional funding for forestry-related projects, specifically for education programs. Grant funds can offset costs for new program initiatives.

Cal Fire is not the only agency that disseminates grant funds. Funding is available through the federal government in the form of the EJ4 Climate grant program which seeks "to foster climate resilience by improving the capacity of communities to prepare for, withstand, respond to, and recover from hazardous events or disturbances related to climate change, which poses risks to human health, the environment, cultural resources, the economy, and quality of life." Growth and maintenance of the urban forest can be considered an important component of climate resilience, as trees provide services such as heat mitigation and carbon storage. Private foundations can also provide funding for tree-related programs in underserved communities facing major threats from climate change, such as East Palo Alto.

Applying for funding for these programs requires staff time to seek opportunities and to coordinate efforts to carry projects from application to completion. It is fair to say that the greater the investment in staffing, the greater the outcomes in funding are likely to be. The City has been successful in partnering with nonprofits and non-governmental organizations. However, implementation of grants regardless of who the applicant is will result in additional need for staffing resources.

Objective 3: Connect with the community around tree stewardship.

Action 7. Design a cohesive and inclusive public outreach program focused on building awareness of the benefits of trees and how and why trees are protected in the City

Key Points

- Engage with local community groups in urban forest stewardship activities

- Partner with Canopy to identify opportunities to expand current community outreach programs

- Provide up-to-date information about tree protections and management on the City website

- Translate tree information into common non-English languages spoken in East Palo Alto

- Designate a forum for the public to engage with tree management and protection actions

- Identify ways for the City to celebrate trees

Community survey results showed less than half of respondents knew that the City protects trees on private and public property. During the public outreach process, community members repeatedly expressed concern over a lack of culturally sensitive materials and a lack of materials available in common languages spoken in the community. These statistics show the need for a City-led outreach campaign to increase awareness of and engagement with the urban forestry program.

A more robust education and outreach program is essential to increasing the community participation needed to reach tree canopy cover goals and to enforce laws that protect trees while also creating an inclusive city that celebrates diversity. Most trees in the city, as well as most opportunities for additional tree planting, are on private property, making private residents a key partner in growing and stewarding the urban forest.

What does a more robust program look like for East Palo Alto? The City could take a four part approach:

1. Engage with active community groups,

2. Enhance informational materials,

3. Create a public forum for community members to be heard on tree issues, and

4. Find ways to celebrate trees.

Engage with community groups

Canopy is currently leading several community programs, including most of the tree planting work occurring in East Palo Alto. As a part of tree planting efforts, Canopy initiated a "tree champions" program, which community members and Canopy staff report has been incredibly successful. Tree champions are residents who are excited about tree planting, who can lead the way and reach out to their neighbors about receiving free trees. These volunteers are involved in planning planting events and recruiting local volunteers. As residents sign up to receive trees, Canopy's trained staff or volunteers visit each home to conduct a site assessment and work directly with each resident to select appropriate planting sites and tree species, coordinate planting day logistics, and offer follow-up tree care information.

Additionally, Canopy is already running and could expand Youth-Led Community Engagement Days and Community Surveys, through a paid internship program with East Palo Alto youth (the Teen Urban Foresters, or TUFs, program). Canopy's paid teen interns design and lead two community engagement days per year, with a goal of gathering survey responses from residents and building greater presence and trust in the community.

Massaging a root ball before planting prevents girdling roots. Photo: Canopy.

Canopy is a valuable partner for East Palo Alto. The City should seek to enhance and grow this partnership. However, the City should also look for opportunities to develop new partnerships with other local organizations. The following groups are essential points of contact for the City:

- Anamatangi Polynesian Voices
- Canopy
- Climate Resilient Communities
- East Palo Alto Community Alliance and Neighborhood Development Organization (EPACANDO)
- East Palo Alto Council of Tenants Education Fund
- El Comite
- Ecumenical Hunger Program
- Faith communities & churches
- Fresh Approach
- Nuestra Casa
- Ravenswood School District
- Ravenswood Business District
- Youth United for Community Action (YUCA)

Providing information to and at community gatherings where these organizations are meeting with staff who are well versed on tree protections and tree canopy goals is an effective strategy to build trust and take action on tree canopy goals. It is an opportunity to educate the community on the benefits of trees, how and why trees are protected, and the role individuals can play in building the urban forest. These opportunities were highlighted by the faith community, whose Sunday services are very well attended. For example, if an East Palo Alto staff member or volunteer were to speak at Sunday services about the need for planting, care and maintenance of trees, and how and when they are protected, this information could reach thousands of residents.

Additionally, there is an opportunity to work with Canopy to host educational opportunities for contractors and landscapers (people who perform work within the dripline of a tree or on a tree) on tree care, the importance of planting, and best management practices. These programs could be designed to be low or no cost to residents. Workshops for community members and landscape professionals on pruning techniques and tree care could be offered by the City in conjunction with Canopy and/or the Cities of Menlo Park and Palo Alto with the goal of raising the level of tree care in the area.

Enhance informational materials

The materials currently available online to provide community members with information about trees and tree protections are lacking. The City website hosts one document pertaining to the tree protection ordinance, which directs community members on how and when trees are protected, how to obtain the necessary materials needed for tree preservation, and how to apply for a tree removal permit. However, the content of this document is different from the body of the tree protection ordinance in Chapter 18.28, and should be corrected to reflect accurate information. Additionally, the document is only available in English. Languages commonly spoken by East Palo Alto residents include Spanish, Tongan, and Samonan. Translating all materials into these languages is crucial for the success of any public outreach campaign.

Information on City trees, the benefits of trees, and the tree preservation ordinances deserves a full webpage on the City website. This page should have information on whom to contact about public tree maintenance and policies, best management practices, tree protection policies, general tree care information, and links to request tree planting from Canopy. The page should also host this Urban Forest Master Plan and any updates on its implementation after adoption. The website should be translated into the common languages spoken in East Palo Alto. Any old materials with inaccurate information should be removed. In the future, content could be added to further engage with website visitors, such as videos in which staff describe the processes of the tree protection ordinance or how to avoid damaging tree pruning practices.

Designate a public forum for tree issues

At community meetings, community members expressed frustration with the lack of opportunities to engage with the City around tree care and protection. These frustrations reflected a lack of communication between the City and the community.

The urban forest management program needs a forum for the public to engage with the City regarding tree issues. The success of the program hinges on how effective it is in reflecting the interests and values of the community. The forum can take the form of a tree board, or a select standing committee with administrative functions such as setting goals, developing policies, and evaluating the success of the program.

One way the interests of the community could lend support and exercise their voice is through the private tree removal permit appeals process. In other cities, public participation is achieved through tree removal permit appeal review and adjudication by the Planning Commission (Atherton) or another advisory board such as a Beautification Advisory Commission (Vallejo) or Heritage Tree Board of Appeals (Pleasanton).

The recommended revisions to the private tree protection ordinance (Action 5) include designating a public body to review tree removal applications as well as any appeals that arise. This body could also serve as a place where community members can voice their concerns and learn about community planting efforts.

Celebrate trees!

Canopy hosts an Arbor Day celebration annually in East Palo Alto. Expanding on this event and adding other tree planting and tree education days to the City calendar will demonstrate the City's commitment to trees. These events don't have to be focused on trees alone, but could also take the form of having a staff person attend a block party or asking Canopy to do a tree planting demonstration at another City function.

Trees are a great conversation starter. Trees are important to the residents of East Palo Alto and can provide a pathway to a more engaged community.

Ravenswood City School District Superintendent Gina Sudaria with Costaño Elementary students. Photo: Canopy.

Action 8. Become a Tree City USA

Key Points
- Following guidance in Action 4, update the public tree ordinance to set standards of care

- Following guidance in Actions 5 and 7, designate or create a public body to review tree removal applications and appeals and to serve as a public forum for tree-related issues

- Following guidance in Action 7, plan celebratory activities for Arbor Day

Community members felt a lack of commitment to the urban forest from the City and wished to see the community more strongly protect and grow trees. In addition to clearly identifying responsibilities and procedures for stewarding public trees within the municipal code, the City should also find ways to show leadership in urban forest stewardship and set an example for the community.

One of the most effective ways for a community to show a dedication to trees is to become a Tree City USA. This program, run by the Arbor Day Foundation, recognizes cities across the country which are investing in and stewarding their urban forests. Most cities near East Palo Alto have acquired Tree City USA status, including Palo Alto, Mountain View, Sunnyvale, Menlo Park, Redwood City, San Carlos, Atherton, Newark, and Fremont.

Four requirements must be met for the City to become a Tree City USA:

- Form a tree board or urban forestry department

- Create a tree care ordinance

- Have a community forestry program with an annual budget of no less than $2 per capita

- Observe Arbor Day

The City would submit an application showing that they meet the qualifications above by the annual deadline each year in order to be approved.

The four requirements to become a Tree City USA will be met if the recommendations from Actions 4, 5, and 7 are followed.

Form a tree board or urban forestry department

Arbor Day recommends establishing a tree board or urban forestry department in order to clearly assign responsibility and accountability for tree work conducted in a city. Currently, the Public Works maintenance division is responsible for tree care, with no designated urban forestry department. If the public tree ordinance is updated to designate the division currently responsible for urban forestry management as an "Urban Forestry" section, then the City will be able to meet this qualification. No change is needed to the work performed.

Additionally, Action 5 of this Plan includes a recommendation for revising the tree protection appeals process, and Action 7 recommends using the forum for this appeals process as an opportunity for public dialogue and oversight of urban forestry activities in the city. Designating a public body such as the Planning Commission to handle all tree-related issues and to serve as a public forum for engagement with urban forestry issues would also meet this requirement.

Create a tree care ordinance

To become a Tree City USA, East Palo Alto will need to update both the public tree ordinance and the private tree ordinance to set tree care standards for trees on public and private property, as described in Actions 4 and 5.

Community forestry program budget

The current tree maintenance budget is about $120,000 annually for contracted tree care operations, plus approximately $30,000 per year for City staff time. The dollars already budgeted more than meet the requirement for $2 dollars per capita on tree care, or roughly $60,000.

Arbor Day celebration

An annual Arbor Day Proclamation is required for the City to qualify as a Tree City. In Action 7, we recommend holding an Arbor Day celebration as a way to positively connect with the local community around trees. Since 2006, Canopy has hosted tree planting events with the City. These events are publicized through the City and in many cases have been recognized as Arbor Day celebrations. These events can be held by Canopy and/or in conjunction with Canopy, continuing the current practice.

Next Steps

The development of this Urban Forest Master Plan is an important step to evaluate where the City is now and how it can work to grow the urban forest. While ideally the City will pursue all actions described above, constraints of budget, time, and staffing will make some actions simpler and faster to achieve than others. The City should regularly revisit this Plan and evaluate progress against the suggested actions. Review should take place as a part of the budget process. Progress will require an investment, but the long term benefits of trees in the City will be realized over time, moving East Palo Alto closer to the community vision of a greener and more resilient future.

Fr. Goode, Mayor Abrica, Assemblymember Marc Berman, and other members of the community at the 2018 MLK Day of Service tree planting at Rich May Memorial Field. Photo: Canopy.

References

Berland, A., Shiflett, S. A., Shuster, W. D., Garmestani, A. S., Goddard, H. C., Herrmann, D. L., & Hopton, M. E. (2017). The role of trees in urban stormwater management. *Landscape and Urban Planning*, *162*, 167–177. https://doi.org/10.1016/j.landurbplan.2017.02.017

City of East Palo Alto. Title 12, Chapter 12.16 - Cutting And Trimming Trees On City Streets And Public Places. Retrieved January 25, 2022 from https://library.municode.com/ca/east_palo_alto/codes/code_of_ordinances?nodeId=TIT12STSIPUPL_CH12.16CUTRTRSTPUPL

City of East Palo Alto. Title 13, Chapter 13.24, Article VII. Section 13.24.410 - Street trees. Retrieved January 25, 2022 from https://library.municode.com/ca/east_palo_alto/codes/code_of_ordinances?nodeId=TIT13PUSE_CH13.24WASY_ARTVIILAIRST_13.24.410STTR

City of East Palo Alto. Title 13, Chapter 13.24, Article VII. Section 13.24.430 - Parking areas. Retrieved January 25, 2022 from https://library.municode.com/ca/east_palo_alto/codes/code_of_ordinances?nodeId=TIT13PUSE_CH13.24WASY_ARTVIILAIRST_13.24.430PAAR

City of East Palo Alto. Title 18, Chapter 18.28 - Landscaping and Trees. Retrieved January 25, 2022 from https://library.municode.com/ca/east_palo_alto/codes/code_of_ordinances?nodeId=EAPAALDECO2018EDCUORNO09-2020ADDE152020_TIT18DECO_ART3REAPALZO_CH18.28LATR

City of Pleasanton. (2020). *City of Pleasanton Shade Tree Guidelines for Commercial Properties.* Retrieved January 19, 2022 from http://www.cityofpleasantonca.gov/civicax/filebank/blobdload.aspx?BlobID=34981

Dahlhausen, J., Biber, P., Rötzer, T., Uhl, E., & Pretzsch, H. (2016). Tree species and their space requirements in six urban environments worldwide. *Forests*, *7*(6). https://doi.org/10.3390/f7060111

International Code Council. (2019). 2019 California Green Building Standards Code California Code of Regulations, Title 24, Part 11. ISBN: 978-1-60983-895-9

Lu, J. W. T., Svendsen, E. S., Campbell, L. K., Greenfeld, J., Braden, J., King, K. L., & Falxa-Raymond, N. (2010). Biological, social, and urban design factors affecting young street tree mortality in New York City. *Cities and the Environment*, *3*(1), 5. https://doi.org/10.1201/b21179-13

McPherson, E. G., & Muchnick, J. (2005). Effects of Shade on Pavement Performance. *Journal of Arboriculture*, *31*(6), 303–310.

Mullaney, J., Lucke, T., & Trueman, S. J. (2015). A review of benefits and challenges in growing street trees in paved urban environments. *Landscape and Urban Planning*, *134*, 157–166. https://doi.org/10.1016/j.landurbplan.2014.10.013

North, E. A., D'Amato, A. W., & Russell, M. B. (2018). Performance metrics for street and park trees in urban forests. *Journal of Forestry*, *116*(6), 547–554. https://doi.org/10.1093/jofore/fvy049

Sanders, J. R., & Grabosky, J. C. (2014). 20 years later: Does reduced soil area change overall tree growth? *Urban Forestry and Urban Greening*, *13*(2), 295–303. https://doi.org/10.1016/j.ufug.2013.12.006

Scott, K. I., Simpson, J. R., & McPherson, E. G. (1999). Effects of tree cover on parking lot microclimate and vehicle emissions. *Journal of Arboriculture*. 25(3), 129-142.

Vogt, J., Hauer, R. J., & Fischer, B. C. (2015). The costs of maintaining and not maintaining the urban forest: a review of the urban forestry and arboriculture literature. *Arboriculture & Urban Forestry*, *41*(6), 293–323.

U.S. Environmental Protection Agency. (2016). Stormwater Trees Technical Memorandum. Tetra Tech, Inc. Retrieved September 20, 2021 from https://www.epa.gov/sites/production/files/2016-11/documents/final_stormwater_trees_technical_memo_508.pdf

RESOURCES

Arbor Day: Information on the Tree City USA program. https://www.arborday.org/programs/treecityusa/

Canopy: Local information on tree planting and care, access to events, tree planting opportunities, and other resources. https://canopy.org/

California ReLeaf: Statewide program support community urban forestry, resources for grants, programs, and information. https://californiareleaf.org/

International Society of Arboriculture: Tree care materials and resources. https://www.treesaregood.org/treeowner

SelecTree: Online tool to identify and search for trees with particular characteristics, to support tree planting. https://selectree.calpoly.edu/

Shade Tree Guidelines, Example from City of Pleasanton: http://www.cityofpleasantonca.gov/documents/landscape/Shade%20Tree%20Guidelines%20for%20Commercial%20Developments%202020.pdf

Tree Care Information (in Spanish): Tree care information and resources from the International Society of Arboriculture available in Spanish. https://www.treesaregood.org/treeowner/spanish

When we plant trees, we plant the seeds of peace and the seeds of hope. Wangari Maathai, 1940-2011. Photo: Canopy.

www.ingramcontent.com/pod-product-compliance
Lightning Source LLC
Chambersburg PA
CBHW061149030426
42335CB00003B/155